Essays in
Biochemistry

volume **26** 1991

Essays in Biochemistry

edited by K.F. Tipton

PORTLAND PRESS

Essays in Biochemistry is published by Portland Press Ltd
on behalf of the Biochemical Society

Portland Press Ltd.
59 Portland Place
London W1N 3AJ
U.K.

British Library Cataloguing-in-Publication Data
Essays in Biochemistry: 26.
    I. Tipton, K.F.
    574.19
ISBN 1-85578-007-0
ISSN 0071-1365

Typeset by Portland Press Ltd and printed in
Great Britain by the University Press, Cambridge

Contents

4 Artificial cell adhesive proteins
Kiyotoshi Sekiguchi, Toshinaga Maeda and Koiti Titani

5 The urea cycle: a two-compartment system
Malcolm Watford

6 Antibody engineering: an overview
Richard O'Kennedy and Paul Roben

Preface

The series of *Essays in Biochemistry* was originally started by Professor Peter Campbell in 1965. After 25 volumes and a succession of editors, it was decided that his original and highly successful concept might be further improved by increasing the number and diversity of chapters. However, the original aim of covering exciting and rapidly developing areas of biochemistry and molecular and cellular biology which would be of particular interest to students and their teachers remains unchanged.

It is hoped that the reader will find the new series useful and stimulating. Any comments, suggestions or criticisms would be most welcome. I would particularly value suggestions for topics and potential authors for inclusion in future editions.

K.F. Tipton
Department of Biochemistry
Trinity College
Dublin 2
Ireland
July 1991

The authors

Steven Rose has been Professor of Biology and Director of the Brain and Behaviour Research Group at the Open University since its foundation in 1969. He studied biochemistry at Cambridge, obtained a Ph.D. in neurochemistry from the Institute of Psychiatry in London in 1962 and has spent postdoctoral periods in Oxford, Rome and Imperial College, London. His laboratory research is focused on the neurobiology of learning and memory, working with the chick model described in this essay. He is also the author of a number of books, including *The Chemistry of Life* (Penguin, new edition published 1991), *The Conscious Brain* (reprinted, Paragon House, 1989) and *Molecules and Minds* (Open University Press, 1988). With Hilary Rose, a sociologist, he has published *Science and Society* (Penguin, 1970) and *The Political Economy of Science* (Macmillan, 1976), and with the geneticist Richard Lewontin and psychologist Leo Kamin *Not in our Genes* (Penguin, 1984). His new book, *Memories are Made of This*, will be published by Bantam later this year.

Edward H. Byard is an Associate Professor in the Biology Department, University of Winnipeg, Winnipeg, Manitoba, Canada. He does research in tubulin molecular biology, primarily in the nematode worm, *Caenorhabditis elegans,* and the protozoan, *Trypanosoma brucei*. This review was written whilst he was visiting Professor Keith Gull's laboratory in the Department of Biochemistry and Molecular Biology, University of Manchester. **Bodo M.H. Lange** is a Ph.D. student in Professor Keith Gull's laboratory in the Department of Biochemistry and Molecular Biology, University of Manchester. For his Diploma thesis he worked on microtubules and microtubule-associated proteins. He is doing his doctoral research on the molecular biology of the centriole.

Bjørn K. Drøbak is in the Department of Cell Biology, John Innes Centre for Plant Science Research, Norwich, where he heads a group involved in research on biochemical and molecular aspects of plant signal perception and transduction. He received his M.Sc. (horticulture) from the Royal Veterinary and Agricultural University of Denmark in 1984, and carried out postgraduate research on calcium and the plant phosphoinositide system at the D.S.I.R. Mount Albert Research Centre, New Zealand, during 1984–1986. From 1987 to 1989 he was a postdoctoral research associate at the University of East Anglia/John Innes Institute. He was appointed to his current position in 1989.

Kiyotoshi Sekiguchi, **Toshinaga Maeda** and **Koiti Titani** have been collaborating since 1988 to engineer artificial cell adhesive proteins modelled after fibronectin, the best characterized cell adhesive protein in the extracellular matrix. **Kiyotoshi Sekiguchi** is associate professor of Fujita Health University School of Medicine. He received his Ph.D. in biochemistry from Osaka University in 1978. **Toshinaga Maeda**

earned his Ph.D. in molecular biology from Niigata University in 1987 and is currently research associate in Osaka Medical Center for Maternal and Child Health Research Institute. **Koiti Titani** is professor of Fujita Health University School of Medicine. He received his Ph.D. from the University of Tokyo, in 1960 and was research professor of biochemistry at University of Washington until 1986.

Malcolm Watford obtained the degree of Bachelor of Science in Applied Biology from Trent Polytechnic in 1974. He then researched for the degree of Doctor of Philosophy in the laboratory of Professor Sir Hans Krebs at Oxford University and did postdoctoral work at the Université de Montréal and Case Western Reserve University. In 1982 he was appointed to a faculty position at Cornell University and since 1990 he has been Associate Professor of Nutritional Sciences at Rutgers University. Since 1988 he has also been Journal Club Correspondent in Metabolic Regulation for *Trends in Biochemical Sciences*. His research interests are the regulation of glutamine metabolism and the molecular biology of the glutaminase isoenzymes.

Richard O'Kennedy obtained his B.Sc. in Biochemistry and Ph.D. at University College Dublin. In 1980 he joined the National Institute for Higher Education, Dublin, as a lecturer. In 1987/88 he was Visiting Scientist at the Department of Tumour Biology, M.D. Anderson Cancer Center, Houston, Texas. At present he is Head of the School of Biological Sciences, Dublin City University. His main research emphasis is on the development of novel antibodies and their applications in analysis. **Paul Roben** is a graduate of Dublin City University where he obtained a B.Sc. degree in Biotechnology, specialising in genetics, process engineering and immunology. He then worked in the biochemistry section of Mitsubishi Petrochemicals Ltd. in Japan. While there, he was involved in polymerase chain reaction research, cDNA library production, DNA synthesis, monoclonal antibody production and related areas. He is now at Dublin City University working on the humanizing of mouse monoclonal antibodies by genetic means.

Marcus Rattray is a lecturer in the Division of Biochemistry at UMDS. He gained his Ph.D. in the department of Biochemistry at Charing Cross and Westminster Medical School in London and became involved in drug abuse research during a postdoctoral fellowship at the National Institutes on Drug Abuse in Baltimore, U.S.A. He is currently researching into the effects of MDMA, and other drugs of abuse, on neuronal gene expression.

Xavier Parés graduated in Biology at the University of Barcelona in 1971. He worked on his Ph.D. (1972–1977) under the supervision of Professor Claudi M. Cuchillo at the Institute of Fundamental Biology and Department of Biochemistry and Molecular Biology in the Autonomous University of Barcelona. From 1978 to 1980 he worked as Research Fellow in Biological Chemistry in Professor Bert L. Vallee's laboratory at Harvard Medical School. Since then he has been Assistant Professor at the Autonomous University of Barcelona. His research interests include the mechanism of RNA hydrolysis by ribonuclease A and the structure and function of alcohol dehydrogenase isoenzymes. **M. Victòria Nogués** received the Degree in Biological Sciences at the Autonomous University of Barcelona in 1976 and, under the tutelage of Professor Claudi M. Cuchillo, she obtained the Ph.D. in 1982 at the same university. At present she is Assistant Professor in Biochemistry and Molecular Biology at that university. She is specially interested in the interaction between RNA and ribonucleases. **Rafael de Llorens** obtained the Degree in Biology (1976) at the Autonomous University of Barcelona. He worked on his Ph.D. (1983) at the same university under the supervision of Professor Claudi M. Cuchillo and Dr. Xavier Parés. From 1983 to 1985 he carried out postdoctoral research, at the Institute of Fundamental Biology, on the synthetic activity of bovine pancreatic ribonuclease A. Since 1985 he has been involved in the study of the use of serum levels of pancreatic ribonuclease as markers of pancreatic disorders, specially pancreatic adenocarcinoma. At present he is Assistant Professor in Biochemistry and Molecular Biology at the Autonomous University of Barcelona. **Claudi M. Cuchillo** graduated in Pharmacy (University of Barcelona) in 1964. He obtained the degree of Doctor in Pharmacy in the same university, under the supervision of Professor V. Villar-Palasí. He then went to the Department of Biochemistry, University College, London, where he worked under the supervision of Professors B.R. Rabin and A.P. Mathias and obtained a Ph.D. degree in 1974. On his return to Spain he was enrolled at the Autonomous University of Barcelona, as assistant professor until 1977 when he became full professor. He is also the Director of the Institute of Fundamental Biology. His main research interest is in enzymology, with special emphasis on ribonucleases.

Ronnitte Badar-Goffer was born in the U.S.A. and moved to Israel when she was 6, where she completed her education with a B.Sc. degree from Bar-Ilan University. She obtained her Ph.D. degree at Cambridge and then helped establish the new NMR Biospectroscopy Group in Nottingham. Her interests are in n.m.r. spectroscopy of the brain with particular emphasis on developing ^{13}C-n.m.r. techniques for elucidating metabolic pathways, as illustrated in her review. **Herman Bachelard** was born and educated in Australia and has worked on the biochemistry of the brain for the past 25 years in the U.K. He has recently moved from being Professor of Biochemistry at the United Medical and Dental Schools of Guys' and St. Thomas's Hospitals in London, to a research professorship in the new NMR Biospectroscopy Centre in Nottingham. There the research concentrates on n.m.r. spectroscopy studies on the brain with emphasis on disorders such as coma, stroke and epilepsy. A past Chief Editor of the *Journal of Neurochemistry*, he has written three textbooks on the biochemistry of the brain and numerous scientific research papers and reviews. He was foundation secretary of the Neurochemical Group of the Biochemical Society and of the European Society for Neurochemistry.

Abbreviations

The abbreviations and conventions used in *Essays in Biochemistry* generally follow those recommended for use in the *Biochemical Journal* [see *Biochem. J.* (1991) **273**, 1–19]. Other abbreviations used in this volume are:

cDNA	complementary DNA
DG	diacylglycerol
G.c.–m.s.	gas chromatography–mass spectrometry
5HT	5-hydroxytryptamine (serotonin)
IMHV	intermediate medial hyperstriatum ventrale
$Ins(1,3,4,5)P_4$	inositol 1,3,4,5-trisphosphate
$Ins(1,4,5)P_3$	inositol 1,4,5-trisphosphate
LPO	lobus parolfactorius
LTP	long term potentiation
MAPs	microtubule-associated proteins
MDMA	methylenedioxymethamphetamine
mRNA	messenger RNA
MTOC	microtubule organizing centre
N.m.r.	nuclear magnetic resonance
N-CAM	neural cell adhesion molecule
NMDA	*N*-methyl-ᴅ-aspartate
PCR	polymerase chain reaction
$PtdIns(4,5)P_2$	phosphatidylinositol 4,5-bisphosphate
$PtdIns4P$	phosphatidylinositol 4-phosphate
TAT	tubulin acetyltransferase
TCP	tubulin carboxypeptidase
TTL	tubulin-tyrosine ligase

The following abbreviations are used for the genetically encoded amino acids:

Alanine	Ala	A
Arginine	Arg	R
Asparagine	Asn	N
Aspartic acid	Asp	D
Aspartic acid or asparagine	Asx	B
Cysteine	Cys	C
Glutamine	Gln	Q
Glutamic acid	Glu	E

Glutamic acid		
or glutamine	Glx	Z
Glycine	Gly	G
Histidine	His	H
Isoleucine	Ile	I
Leucine	Leu	L
Lysine	Lys	K
Methionine	Met	M
Phenylalanine	Phe	F
Proline	Pro	P
Serine	Ser	S
Threonine	Thr	T
Tryptophan	Trp	W
Tyrosine	Tyr	Y
Unknown		
or "other"	Xaa	X
Valine	Val	V

The following symbols are used for nucleotides in nucleic acid sequences (both DNA and RNA; note that the symbol T is used at all positions where U might appear in the RNA):

G	guanine
A	adenine
T	thymine
C	cytosine
N	"any" or "unknown"

1

The biochemistry of memory

Steven P.R. Rose

Brain and Behaviour Research Group, The Open University,
Milton Keynes MK7 6AA, U.K.

WHAT IS MEMORY?

At first sight the title of this Essay may appear paradoxical. What has biochemistry, that most materially grounded of the biological sciences, to do with memory, a seemingly elusive subjective phenomenon which we all experience but may find it hard to define? Indeed for a long time biochemists tended to shy away from the problem on the grounds that it was indeed too difficult or too controversial. Today however the field has come of age; it is the subject of active study in many biochemistry and molecular biology laboratories across the world, and has been singled out for special mention within national and international research programmes, from the US National Institutes of Health's 1990s "Decade of the Brain" to the Japanese Human Frontier Research Programme. The belief that biochemistry is contributing to an understanding of one of the most fascinating and important of all human characteristics provides a strong motivation for those researching on memory.

One of the most striking features of memory is its very durability; eighty- or ninety-year-olds can remember their childhood experiences, retained over a lifetime during which every molecule of their body has been turned over many times. It was realized as long ago as the end of the last century that such stability must mean that when memories are formed, then there would be structural changes in the brain, an alteration in the relationships between its nerve cells (neurons) which in some way could "represent" the memory. However, not all memorized items are retained so long. Presented with a string of seven numbers at the rate of one a second, and immediately asked to repeat them back, most people can manage. If they are asked again half an hour later, most fail. However, if they are told that the numbers are important to remember — say a telephone number — then they are much more likely to succeed, and once they have crossed this half hour barrier, are likely to be able to remember the number for some days at least.

This type of study led to the suggestion that there are at least two stages of memory formation, an unstable, transient short-term phase and a more stable long-term phase. Observations on patients who were recovering from concussion or from the effects of electroconvulsive shock treatments also showed a characteristic loss of memory for events in the hour or so prior to the shock but rather little impairment of earlier memories. Again, this lends credence to the view that while short-term phases of memory may be held by transient changes in the firing properties of particular neurons, longer term memory must require more substantial modification of cells and their connections.

During the first half of this century, and beginning with Pavlov, psychologists began the systematic study of learning and memory in experimental animals. Such studies showed that, after a sufficient number of repetitions of a particular experience, animals could form an association between previously disconnected causes and effects — for instance, that the ringing of a bell presaged the arrival of food (and hence produced salivation), or that the pressing of a lever would be followed by a reward. The former, when an animal does not control its environment, is Pavlovian or classical conditioning; the latter, in which an animal acts on its environment, is Skinnerian or operant conditioning. These are not the only forms of learning; for instance, when an animal learns to run a maze it forms some sort of "cognitive map", an internal model which it uses to orient itself in relation to external cues.

The results of all these forms of learning are changes in the behaviour of the animal which may last days, weeks or longer. However, there are also shorter-term processes which can be regarded as forms of memory. Take an actively exploring snail, and lightly touch one of its eye-stalks with a glass rod; it will instantly retract and cautiously extend it again after a few minutes. Touch it again and the procedure will be repeated, but if the exercise is conducted enough times the snail will become indifferent and no longer respond to the touch of the rod. This is habituation. It is not however a passive process due to biochemical or physiological exhaustion, because if the touch is repeated but is now accompanied by an unfamiliar additional stimulus — a shock to the snail's foot, or a bright light — then the retraction will again occur. This is dishabituation or sensitization.

HEBB SYNAPSES

Recognition that associative learning must involve structural changes in the brain led the psychologist Hebb, in 1949[1], to propose a model of the cellular mechanism of learning which remains the basis for much present day thinking. Consider a neuron, C, on which two neural pathways converge. Suppose that the first, A, is a strong pathway, so that when A fires C is always activated, whereas the second, B, is a weak synaptic connection (see Figure 1). Suppose further, that if A and B fire simultaneously, the activation of C by A results in some retrograde signal to B, so as to strengthen that synaptic connection, then C will have "learned" to fire in response to B as well as A (to give an example, suppose C is a neuron responsible for salivation,

Figure 1. Schematic drawing of Hebb's proposed mechanism for memory formation

"Let us assume then that the persistence or repetition of a reverberatory activity (or "trace") tends to induce lasting cellular changes that add to its stability. The assumption can be precisely stated as follows: *When an axon of cell A is near enough to excite a cell B and repeatedly or persistently takes part in firing it, some growth process or metabolic change takes place in one or both cells such that A's efficiency, as one of the cells firing B, is increased.* The most obvious and I believe much the most probable suggestion concerning the way in which one cell could become more capable of firing another is that synaptic knobs develop and increase the area of contact between the afferent axon and efferent [cell body]. There is certainly no direct evidence that this is so.....There are several considerations, however, that make the growth of synaptic knobs a plausible perception." (*D. O. Hebb: The Organisation of Behavior, Wiley, 1949, pp. 62–63*)

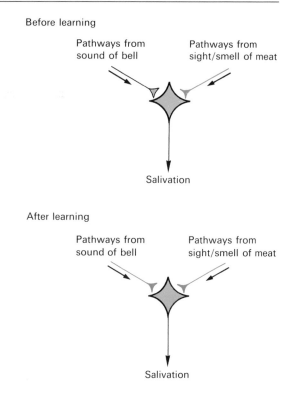

Before learning

Pathways from sound of bell

Pathways from sight/smell of meat

Salivation

After learning

Pathways from sound of bell

Pathways from sight/smell of meat

Salivation

A is a pathway from the olfactory system which fires C on the smell of food, and B is a weak pathway from the auditory system which responds to the sound of a bell, then such a mechanism would result in the animal now responding to the sound of the bell by salivating). The most obvious way that such strengthening of a connection could occur, said Hebb, would be if there was some growth or modification of the synaptic bouton connecting B and C.

A number of variants on such "Hebb synapses" have been proposed in recent years, and they have been the subject of many attempts at neural modelling; in particular, theoreticians working with new types of computing based on parallel distributed architectures have envisaged the brain as containing many small ensembles of neurons, connected in a network of Hebb synapses, which can be strengthened or weakened in order to encode for new information, or to provide, as the current jargon has it, "neural representations" of the external world[2].

STRUCTURAL CHANGES

Nor are such synaptic changes merely theoretical; they can be observed directly at both light and electron microscope level. Figure 2 shows a Golgi-stained preparation of a neuron. This staining technique is capricious in that it only picks out a small proportion of the neurons in any section but it stains each one in its entirety, showing the multiply branching dendrites, studded with spines. Each spine receives a synaptic

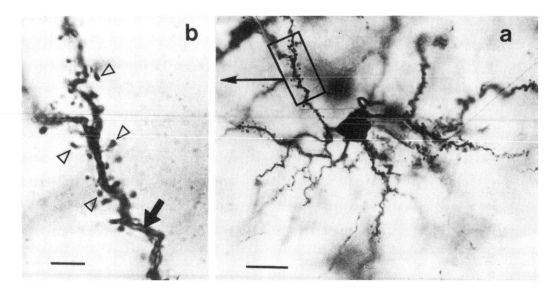

Figure 2. (a) Photomicrograph of Golgi-impregnated large spiny multipolar projection neuron from 2-day old chick IMHV (scale bar = 15 μm) and (b) higher power micrograph of dendritic segment showing dendritic shaft (large arrow) and spines (arrow-heads) (scale bar = 4 μm)
Courtesy Dr. M.G. Stewart.

terminal from presynaptic cells not visible by this staining technique. It has been know for some years that rearing rats in so-called "enriched" versus "deprived" conditions can result in substantial increases in dendritic branching and in the density of spines (that is, numbers of spines per of dendrite) on cortical neurons. Recently it has been found that a single "learning experience" can have similar effects. Figure 2 is of a neuron from a region of the left hemisphere of the chick brain known to be involved in memory formation. The day-old chick has been trained on a simple learning task in which it is offered a small shiny bead to peck; the bead is coated in a distasteful, bitter-tasting substance and as a result of a single peck the chick will avoid a similar bead when offered thereafter. Twentyfour hours after such an experience, there is a 60% increase in the density of the dendritic spines on the neuron, compared with that in a control animal which has pecked only a water-covered bead, or the equivalent right hemisphere region in the trained bird[3].

Such an increase implies either the formation of new synapses or the relocation of existing ones, perhaps from the shafts of the dendrites to the spines — a relocation which, because of its geometrical effect, would have the required Hebb-type consequence of making the synapse more effective. Changes in the actual numbers and dimensions of synapses can be seen in the same region of the chick brain. The electron micrograph of the synapse in Figure 3 shows some of the features that can be measured. Training the chick on the pecking task results in increases in the length of the synaptic apposition zone (the active region between the pre- and postsynaptic sides of the synapse, packed, on the postsynaptic side, with receptors for the transmitters released from the presynaptic side). There are some changes in synaptic dimensions and in one brain region an actual increase in numbers of synapses, but the most striking change is a 60% increase in the numbers of synaptic vesicles which contain the neuro-transmitters[4].

BIOCHEMICAL APPROACHES

Such morphological changes, which have been found in other learning tasks as well, are striking vindications of the idea that learning and memory formation involve some structural synaptic reorganization, but what have they to do with biochemistry? In the 1960s there was a brief flurry of enthusiasm amongst biochemists and molecular biologists for the idea that somehow memories would be coded in the brain in term of specific unique protein, RNA or even DNA molecules. However, the improbability of this idea (and the implausibility of some of the experiments claimed to support it) soon led to its dismissal in respectable biochemical circles[5]. Today, it is recognized that the "biochemistry of memory" is not concerned with the search for unique bio-chemical processes or molecules, but rather with the identification of the sequence of biochemical processes that must be involved in the short- and long-term modifi-cation of synaptic structures and connectivity.

Two approaches to such a search are in principle possible, sometimes called "corre-lative" and "interventive." In the former, an animal is trained on a particular task and a search made for the biochemical changes that accompany the learning and memory formation. The problems with this approach are first that changes are likely to be small and localized, so it is desirable to have some idea in advance of which brain regions may be involved; and second that all measurable learning tasks involve the animal in some type of motor or sensory activity, possibly stress and certainly arousal; it is necessary to devise controls to ensure that observed changes are not merely the sequelae of these necessary concomitants of memory[6].

Figure 3. Electron micrograph showing synapses from the chick IMHV
Interior of synaptic bouton is packed with vesicles (ves). Asterisk shows synaptic contact of bouton with dendritic spine head. Postsynaptic thickening (pst) indi-cated by arrowheads (scale bar = 0.2 µm). Courtesy Dr. M.G. Stewart.

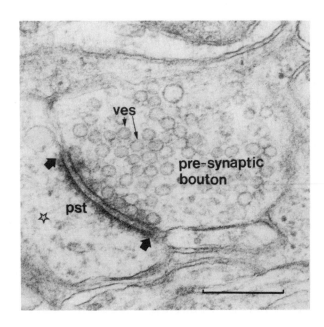

The second, interventive, approach endeavours to disrupt the process of memory formation by blocking a specific biochemical process, for instance with a transmitter antagonist or an enzyme inhibitor. The best known general example of this is the almost universal finding that administration of protein synthesis inhibitors, such as the antibiotics cycloheximide or anisomycin, just before or within an hour after training results in amnesia for the task[7], without affecting recall of already established memories. This should indicate that the blocked biochemical process was necessary for the memory to be formed. Problems with this approach include the need to ensure that the inhibitor does not have other, unwanted effects either on biochemistry or behaviour — for instance, protein synthesis inhibitors cause accumulation of intra-cellular pools of amino acids, which can be both transmitters or neurotoxins; alternatively the inhibitor could cause motor or sensory deficits preventing the animal from actually carrying out the task.

SEA-SLUGS AND FLIES

Whichever approach is adopted, it still necessary to find appropriate "animal models" in which memory can be studied, and these may not be the same as those used in traditional learning psychology, because training adult rodents to run mazes or press levers may require only relatively small-scale synaptic changes, likely to go undetected against the general level of cellular activity. Four such models, two invertebrate, two vertebrate, have proved particularly interesting in the last few years.

The best known of the invertebrates is the marine mollusc *Aplysia californica*, which has been extensively studied by Kandel and his colleagues over the past two decades[8,9]. *Aplysia* has a simple behavioural repertoire and a nervous system comprised in part of large and readily identifiable neurons (sometimes misleadingly referred to as a "simple nervous system"). In particular a number of bodily responses, notably the reflex withdrawal of its gill and siphon when touched by a rod or a jet of water, show habituation, sensitization and even forms of associative learning. The sets of neurons controlling this reflex can be dissected out and studied in isolation. Amongst them are a particular motor neuron with its associated sensory synapse. This single synapse preparation can itself be maintained in culture, giving rise to what Kandel has called "memory in a dish". Although only a fraction of the entire system, this preparation has enabled him to analyse the neural analogues of behavioural phenomena. For instance, during habituation there is a decreased release of the neurotransmitter serotonin from the presynaptic terminal, during sensitization an increase. This altered release is associated with the opening of calcium channels in the membrane, and the triggering of a second messenger cascade involving cyclic AMP and a number of associated protein kinases. The further biochemical details will be discussed in the general schema below.

A second interesting approach has been the search for learning and memory mutants in that favoured organism of geneticists, the fruitfly *Drosophila melanogaster*. *Drosophila* can be taught a number of simple tasks, notably to avoid flying towards particular odours which are associated with electric shock. After subjecting a fly population to a mutagen, flies can be discovered which, whilst not showing any other apparent behavioural deficits, cannot remember the odour/shock association. Several such mutant populations, graced with names like *dunce, turnip* and *rutabaga*, have been identified. Interestingly the biochemical effect of the mutations seems to be directed at sites within the cyclic AMP/ protein kinase complex[10].

THE HIPPOCAMPUS

One of the most popular vertebrate systems has been the rodent hippocampus. This layered neuronal structure (Figure 4) has long been of interest to neurobiologists because of the evidence of its involvement in both human and animal memory processing. Humans with hippocampal damage are able to recover memories laid down before the damage occurred, and show a limited short-term memory capacity. However, they cannot transfer memories from short- to long-term store, and hence they deny knowledge of experiences occurring more than a few minutes previously. Rodents with hippocampal damage lack the capacity to create "cognitive maps" and cannot navigate in an environment in which they are dependent on external cues. In 1973 Bliss & Lømo[11] showed that if the input pathway to a specific hippocampal layer was triggered by a train of high frequency pulses, there was a large potentiation of the firing of the hippocampal cells that persisted for many weeks — a phenomenon they called long term potentiation (LTP). LTP was instantly interesting to neurophysiologists, because it is a physiologically generated stable change in neural properties; in this it might be considered a *model* for memory. But because of the special role of the hippocampus, might it not also prove to be a *mechanism* for memory as well?

Studies of the biochemical processes involved in LTP have focused, first on the transmitters involved, and second on whether the changes which account for the altered properties of the hippocampal cells are pre- or postsynaptic, or some combination of the two[12,13]. There is general agreement that during LTP there is enhanced release of glutamate from presynaptic terminals and that this interacts with upregulated receptors of one glutamate subtype, NMDA (*N*-methyl-D-aspartic acid). Proponents of presynaptic models argue that the next step is a release of some retrograde messenger (candidates include components of the phosphoinositide cascade such as

Figure 4. The mammalian hippocampus (redrawn from Dudai[10])
The diagram shows the location of the hippocampus in the rabbit brain, and, enlarged, a section showing how a hippocampal slice, which can sustain long term potentiation, can be cut. CA1 and CA3 are pyramidal cell fields of the hippocampus; comm, commissural projection to area CA1; DG, dentate gyrus; fim, fimbria; mf, mossy fibres; pp, perforant pathway; Sch, Schaffer collaterals.

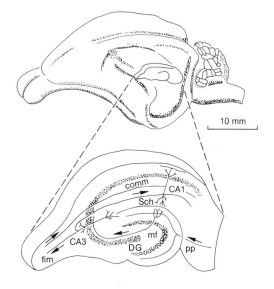

arachidonic acid, nitric oxide or even free radicals), followed by changes at the pre-synaptic membrane. In particular, there is a specific presynaptic membrane phospho-protein, molecular mass around 50kDa, known by different laboratories as B50, GAP43 or F1. The phosphorylation of this protein is dependent on protein kinase C, and changing its phosphorylation state is supposed to affect calcium flow into the cell and hence activate intracellular second messengers. Protein kinase C inhibitors will block LTP. However, adherents to postsynaptic models argue that intracellular injection of specific protein kinase C inhibitors into the postsynaptic cell will also block LTP.

CHICKS

The vertebrate model described earlier, passive avoidance learning, in which a young chick learns to suppress pecking at a bitter tasting bead, has a number of biochemical and behavioural advantages. Training is quick, easy and reliable; the chick has a large and accessible brain and an unossified skull, which means that it is simple to

Table 1. Cellular cascade during memory formation for passive avoidance
Arrows indicate increase in quantity, activity or turnover; locations are to the left or right IMHV or LPO where known. The inhibitors producing amnesia are defined in the text.

Timescale	Process	Location	Inhibited by (amnesia)
1. Seconds to minutes	(a) Glucose uptake ↑	IMHV, LPO (left and right)	
	(b) Receptor binding ↑	IMHV ?	APV, MK801
2. Minutes to hours	(a) Protein kinase C translocation ↑	Left IMHV	
	(b) Presynaptic B50 phosphorylation ↑	Left IMHV	Polymixin B, melittin, H7, staurosporine
	(c) c-fos, c-jun expression ↑	Left IMHV	
3. 1–6 hours	(a) Protein synthesis (tubulin) ↑	IMHV	Cycloheximide, anisomycin, etc.
	(b) Glycoprotein synthesis (pre- and post-synaptic) ↑	Left IMHV, left and right LPO	2-Deoxygalactose, electroshock
	(c) Fucokinase ↑	Right LPO	
	(d) Neuronal bursting ↑	Left and right IMHV, left and right LPO	Electroshock
4. 12–24 hours	(a) Dendritic branching ↑	Left IMHV	Electroshock
	(b) Spine head diameter ↑	Left IMHV	
	(c) Synapse number ↑	Left and right LPO	
	(d) Vesicle number ↑	Left IMHV, left LPO	
	(e) Post-synaptic density length ↑	Left IMHV, left LPO	

inject precursors and drugs. Learning to suppress pecking at the bitter bead initiates a biochemical cascade, which, beginning with pre- and postsynaptic membrane transients and proceeding by way of genomic activation to the lasting structural modification of these membranes, occurs in identified regions of the chick forebrain[14]. These synaptic modifications must form in some way the neural representations of the aversive bead-pecking experience and encode the instructions for the changed behaviour (avoid pecking a bead of these characteristics) that follows. The biochemical, physiological and structural cascade, as it has been identified in the chick, but which bears many resemblances to mechanisms found in other learning models, such as LTP, is shown in Table 1.

The brain regions which showed enhanced neural activity as a result of training were first identified on the assumption that such increased activity would demand greater glucose utilization, detectable using radioactively labelled 2-deoxyglucose, which accumulates as 2-deoxyglucose 6-phosphate in cells at a rate proportional to their glucose utilization. Increases in two regions, the intermediate medial hyperstriatum ventrale (IMHV) and lobus parolfactorius (LPO), could be detected for at least half an hour following pecking at the bitter bead, and the later biochemical, physiological and structural changes were all localized to these regions (Figure 5).

By analogy with LTP, soon after training on the passive avoidance task there is an upregulation both of glutamate (NMDA) and acetylcholine (muscarinic) receptors; NMDA receptor blockers (such as the noncompetitive antagonists MK801 and APV) produce amnesia, and chicks which have learned to avoid the bitter bead peck it instead. Half an hour after training, there is a change in the phosphorylation of the presynaptic B50 protein, seemingly regulated by the translocation of cytosolic protein kinase C to the synaptic membrane. As would be expected, inhibitors of protein kinase (such as melittin, staurosporine or H-7) result in amnesia for the passive avoidance.

LONG-TERM CHANGES

All the processes so far discussed have involved transient effects, receptor regulation, opening of membrane channels and initiation of second messenger cascades. But long term memory must involve more permanent changes to synaptic membrane structure, which requires the synthesis of new membrane proteins. The molecular biological mechanisms involved in triggering the synthesis of such proteins are assumed to involve the initial activation of members of the family of immediate early genes of which the protein oncogenes c-*fos* and c-*jun* are amongst the best-known. c-*fos* and c-*jun* expression is believed to be initiated by signals emanating from the membrane, especially the opening of calcium channels and the activation of the phosphoinositide cycle mediated by the phosphorylation steps described above[15]. At 30 minutes after training chicks on the passive avoidance task, c-*fos* and c-*jun* mRNAs are expressed in IMHV and LPO. Although the induction of these genes is notoriously sensitive to many types of sensory stimulation (including pecking at a water-coated bead) it has been possible to show that these increases are directly associated with learning the new task[16].

Whatever the intervening intracellular signals and genomic mechanisms, within an hour after training there is enhanced synthesis of a variety of proteins intended for

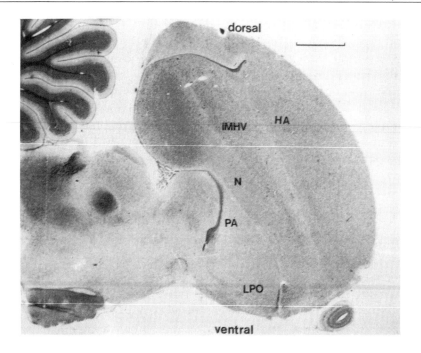

Figure 5. Saggital section of 1-day old chick brain, nissl stained
HA, hyperstriatum accessorium; IMHV, intermediate medial hyperstriatum ventrale; N, neo-striatum; PA, paleostriatum augmentatum; LPO, lobus parolfactorius (scale bar = 1.5 mm).

export from the neuronal cell body. Particularly relevant here are the glycoproteins of the synaptic membrane, because of the major role that several glycoprotein families (for instance, the family of neural cell adhesion molecules or N-CAMs[17]) play in intracellular recognition and in stabilizing intercellular connections, especially during neural development. For many hours after training chicks, there is enhanced incorporation of radioactively labelled glycoprotein precursors such as fucose into both pre- and post-synaptic membrane glycoproteins, regulated by increased activity of the rate-limiting enzyme fucokinase. A number of glycoproteins are involved, including a presynaptic component of molecular mass around 50kDa and postsynaptic components of molecular masses 100–120 kDa and 150–180 kDa; these molecular masses are interestingly close to those of the N-CAMs[18]. As with the phosphorylation and protein synthesis steps of this series of reactions, it would be expected that, if the synthesis of glycoproteins was a necessary step in the formation of long-term memory, then inhibiting this synthesis should produce amnesia. Fortunately there is a specific inhibitor of fucoglycoprotein synthesis, 2-deoxygalactose, a competitive inhibitor to galactose, which, incorporated into the nascent glycoprotein chain, prevents terminal fucosylation. In chicks, as well as in rats, 2-deoxygalactose injected within an hour of the time of training produces amnesia for the task in animals tested 24 hours later.

It is presumed that the synthesis of the new membrane glycoproteins results in the pre- and postsynaptic structural changes that can be observed in the Golgi-stained and electron micrograph pictures. If the Hebb model is correct, such changes must alter the physiological connectivity of the neurons, and indeed recording from IMHV and LPO in the hours after training reveals substantial increases in the spontaneous high-frequency neural activity ("bursting") which is rather reminiscent of LTP[14].

MAKING MEMORIES

What these studies — which are still very much in progress — reveal is at the least a plausible biochemical mechanism by which changes in neuronal connectivity that constitute the neural representation of associative memories can be formed. What form might such a neural representation take? It is not clear if there is but one general biochemical cascade involved in all forms of learning, or many distinct mechanisms will distinguish, say association learning in *Aplysia* from maze learning in a rat or the multiple facets of human memory. If past experience of the parsimony principle by which biochemistry seems to operate are a guide, we may find both broad universals and fascinating uniqueness. But, as a casebook example, consider the chick learning the passive avoidance task. The default response to a bead, so to say, must be to register the presence of an object of particular size and colour, and to activate the motor responses of pecking towards it. Experiencing the taste of the bitter bead must activate the biochemical machinery that will create in the brain some sort of model against which any fresh presentation of a bead can be matched. If the match is good, then the output response of pecking is inhibited. Although the initial storage of the model might involve sets of cells and synapses localized in the IMHV, it is probable that subsequently the bird's concept of "bitter bead" is classified in several different ways, in multiple sites, and involving much greater economy of cellular machinery than implied by the large changes described above.

The ultimate task of neuroscience is to discover how such cellular phenomena "translate" into the experiences of conscious behaviour and memory; biochemistry has a part to play in uncovering such translation rules, but only if we always remember that it provides only part of the much larger picture which the myriad complex interactions of the neurons and synapses of the brain make possible.

REFERENCES

1. Hebb, D.O. (1949) *The Organisation of Behavior*, Wiley, New York

2. Rumelhardt, D.E., McClellan, J.L. and PDP research group (1987) *Parallel Distributed Processing: Explorations in the Microstructure of Cognition*, 2 vols., MIT Press, Cambridge, MA

3. Patel, S.N. & Stewart, M.G. (1988) Changes in the number and structure of dendritic spines 25hr after passive avoidance training in the domestic chick, *Gallus domesticus. Brain Res.* **449**, 34–46

4. Stewart, M.G. (1991) in *Neural and Behavioural Plasticity; the Use of the Domestic Chick as a Model* (Andrew, R.J., ed.), Oxford University Press, Oxford, in the press

5. Rose, S.P.R. (1992) *Memories are Made of This*, Bantam Books, New York and London

6. Rose, S.P.R. (1981) What should a biochemistry of learning and memory be about? *Neuroscience* **6**, 811–822

7. Davis, H.P. & Squire, L.R. (1984) Protein synthesis and memory: a review. *Psychol. Bull.* **96**, 518–559

8. Hawkins, R.D. & Kandel, E.R. (1984) Is there a cell-biological alphabet for simple forms of learning? *Psychol. Rev.* **91**, 375–391

9. Goelet, P., Castelucci, V.F., Schacher, S. & Kandel, E.R. (1986) The long and the short of long-term memory – a molecular framework. *Nature (London)* **322**, 419–422

10. Dudai, Y. (1989) *The Neurobiology of Memory,* Oxford University Press, Oxford

11. Bliss, T.V.P. & Lømo, T. (1973) Long-lasting potentiation of synaptic transmission in the dentate area of the unanaesthetized rabbit following stimulation of the perforant path. *J. Physiol. (London)* **232**, 331–356

12. Lynch, G. & Baudry, M. (1984) The biochemistry of memory – a new and specific hypothesis. *Science* **224**, 1057–1063

13. Bliss, T.V.P. (1990) Memory: maintenance is presynaptic. *Nature (London)* **344**, 698–699

14. Rose, S.P.R. (1991) in *Neural and Behavioural Plasticity; the Use of the Domestic Chick as a Model* (Andrew, R.J., ed.), Oxford University Press, Oxford, in the press

15. Chiarugi, V.P., Ruggiero, M. & Coradetti, R. (1989) Oncogenes, protein kinase C, neuronal differentiation and memory. *Neurochem. Int.* **14**, 1–9

16. Anokhin, K.V. & Rose, S.P.R. (1991) Learning-induced increase of immediate early gene messenger RNA in the chick forebrain. *Eur. J. Neurosci.* **3**, 162–167

17. Edelman, G. (1987) *Neural Darwinism*, Basic Books, New York

18. Rose, S.P.R. (1989) Glycoprotein synthesis and post-synaptic remodelling in long-term memory. *Neurochem. Int.* **14**, 299–307

2

Tubulin and microtubules

Edward H. Byard* and Bodo M. H. Lange

Department of Biochemistry and Molecular Biology, University of Manchester, Manchester M13 9PT, U.K.

INTRODUCTION

Most textbooks of cell biology describe, in an introductory chapter, the "typical" cell. This cell is nearly always a non-descript, spherical eukaryote (and usually an animal cell) with a bland-looking interphase nucleus. The "typical" cast of organelles is present, and is clearly enough labelled — the endoplasmic reticulum, the Golgi complex, the ribosomes, the plasma membrane, the mitochondria. If this picture truly describes cells as they really exist, students of cell biology could be forgiven for closing the book, and enrolling in physics instead!

The reality, of course, is much different. Cells are rarely spherical, but instead exhibit a constellation of shapes and sizes, from filamentous to fubsy, from massive to minute. Some live a solitary existence, whilst others are pushed, pulled and shaped by other cells around them. And cells move. Single cells move from place to place with purpose and direction. Organelles are moved within the cytoplasm of cells, often over long distances, and to specific destinations. Chromosomes are replicated, and the copies are carefully shepherded into daughter cells during cell division.

Cells maintain their shape and sustain their necessary movements by the workings of the various components of the *cytoskeleton*, a complex network of filaments and tubules that form an intracellular scaffolding, that has the added property of being able to be torn down and dynamically remodelled. There are three major components of this scaffolding (which, incidentally, are often not even labelled in those pictures of the "typical" cell mentioned above!): *microfilaments, intermediate filaments* and *microtubules*. The study of the molecular biology of these components in cells is an important

---

*Permanent address: Department of Biology, University of Winnipeg, Winnipeg, Manitoba, Canada R3B 2E9.

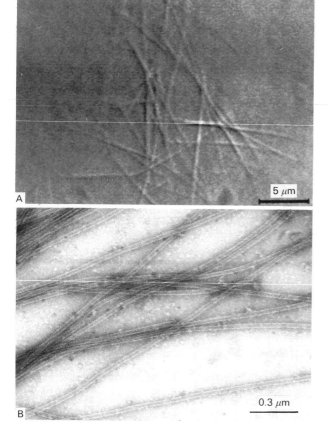

Figure 1. Isolated microtubules, in this case prepared by assembly *in vitro*, can be visualized using the light microscope (A) or the electron microscope (B)

The image shown in (A) was produced by using a video enhancement technique. Unfixed and unstained microtubules are visible as a meshwork; this method allows properties of microtubules to be studied *in vitro*. In (B), a similar preparation has been negatively stained; that is, the microtubules are outlined by the stain, and the greater resolving power of the electron microscope allows some internal detail of protofilaments to be seen.

and active area of cell biology research, and in the few pages which follow we describe some of this research as it pertains to the *microtubules*, structures which are built from simple precursors into an apparently simple three-dimensional polymer, but which appear to have, within their chemistry and structure, the potential to form spatially complex and functionally regulated networks. We do not intend to provide a full review of the work in this field (since many excellent reviews already exist, and are cited in the text), but rather to highlight some of the areas of active research.

TUBULIN AND THE ASSEMBLY OF MICROTUBULES

α- and β-tubulins

In eukaryotic cells, there are two major tubulin polypeptides present, which are called α-tubulin and β-tubulin. These polypeptides are globular, soluble, and very similar to one another. Each of the polypeptide monomers has a molecular mass of about 55 kDa, and the native form of tubulin in cells is an α–β heterodimer, which can self-assemble into the insoluble, polymerized microtubule. This property of self-assembly, which can be accomplished *in vitro*, has been used extensively in studies of the detailed chemistry of the tubulins. If the concentration of the tubulin heterodimers is maintained above a critical concentration, spontaneous assembly of microtubules will occur (Figure 1). The essential requirements of this assembly are that the temperature is between 25 and 37 °C, and that the nucleotide GTP is present at about 1–2 mM. Since microtubules are insoluble, they can be collected by centrifugation. Cooling the solutions leads to microtubule disassembly, so "cycles" of warming and cooling yield highly enriched tubulin. These preparations have allowed some detailed analysis of the tubulin molecule itself, as well as the kinetics of assembly (see the recent review by Mandelkow & Mandelkow for a discussion of these aspects[1]).

Very recently, a third, divergent, tubulin called *γ-tubulin* has been identified[2]. Very little is known, as yet, about its biochemistry, or its assembly properties, although it is clear that it does not occur in large amounts in microtubules, being restricted largely to the ends of the microtubules, particularly those near microtubule organizing centres[3]. Some early, but as yet unpublished, results suggest that γ-tubulin is present in many organisms, and thus may represent another important addition to the tubulin family.

Assembly of microtubules

In vitro assembly has been useful in addressing some general questions in relation to microtubule assembly, but more work is now aimed at understanding the dynamics of assembly *in vivo*, since the *in vitro* model is inadequate in some aspects. It is possible, *in vitro*, to visualize that an assembled microtubule consists of rows of tubulin molecules called *protofilaments*, which are, in turn, folded up into a tube of about 25 nm outer diameter. There are normally 13 protofilaments in a microtubule but the number of protofilaments is variable, both *in vitro* and *in vivo*: some organisms have microtubules with 11 or 15 protofilaments. In addition, the length of a microtubule *in vitro* appears to be indeterminate, whereas the cell obviously has rather precise control over microtubule length. Much effort is now focused on detailed analysis of how microtubules are assembled *in vivo*, and this effort is beginning to bear fruit.

For example, it is clear that the assembled microtubule has a distinct polarity, which probably influences the way in which microtubules grow, and are oriented in cells. In the first instance, the orientation of the association between heterodimers is such that the α end of one heterodimer associates with the β end of the next. Thus, one end of a microtubule has free α-tubulins, whereas the other has free β-tubulins. As yet, it is not known *in vivo* how these "ends" relate to the arrangement of microtubules in cells, but the answer to the question of which end of a microtubule is which could be interesting.

Secondly, the rate of heterodimer addition at one end of a growing microtubule is usually faster than at the other end. The fast-growing end is termed the "plus" end, and the slow-growing end the "minus" end. It is not completely clear how this differential rate of assembly is accomplished, but a recent view, proposed primarily by Kirschner and his colleagues[4], suggests that the binding of the nucleotide GTP may be a central factor. As mentioned earlier, microtubule assembly has a requirement for the nucleotide GTP, which is noncovalently bound to tubulin on both the α- and β-subunits. At the β-site, GTP is hydrolysed to GDP at a rate which is much slower than the rate of the assembly of the heterodimers into a microtubule at the "plus" end. This means that, at one end of the microtubule, there is a "cap" of GTP–tubulin, which is of a significant size when microtubules are being rapidly assembled. After

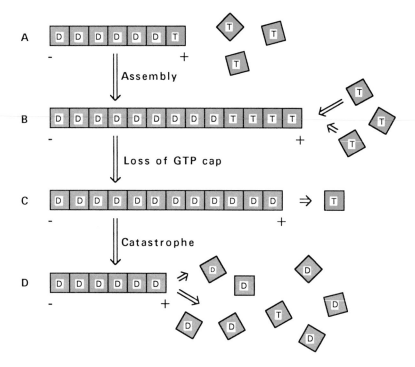

Figure 2. A schematic version of dynamic instability in microtubules
(A) Microtubule growth begins from a nucleation site, in this case a partially assembled microtubule. The tubulin heterodimers (shown here as blocks for simplicity) have either GDP (labelled D, coloured grey) or GTP (labelled T, coloured pink) bound to them. The soluble heterodimers are GTP–tubulin. (B) If the concentration of soluble heterodimers is sufficient, microtubule assembly proceeds, at the "+" end of the tubule. Initially, GTP–tubulin attaches to the growing end; after a time, the GTP is hydrolysed to GDP. (C) If conditions in the cell are such that the concentration of GTP–tubulin drops, the GDP–tubulin is the predominant form of tubulin in the tubule. (D) GDP–tubulin readily detaches from the "+" end of the tubule in the absence of GTP–tubulin, and rapid depolymerization occurs. Long tubules will shorten rapidly; short tubules will disappear altogether ("catastrophe"). This process can be reversed ("rescue") if the concentration of free heterodimers is increased.

the hydrolysis of GTP to GDP at the exchangeable site on β-tubulin, the GDP–tubulin is more readily disassembled from an assembled microtubule. Thus the size of GTP caps may well determine the rates of microtubule assembly and disassembly; if the GTP cap is small, or even absent altogether, the microtubule will disassemble rapidly (and perhaps even catastrophically!). This so-called *dynamic instability* may well explain how microtubules *in vivo* can be rapidly remodelled (Figure 2).

GENETICS OF TUBULIN

Tubulin genes

The intensive application of the techniques of molecular biology has, very rapidly, allowed an extensive characterization of the nature and occurrence of tubulin genes in a wide variety of organisms. In the ten years since Cleveland and his colleagues isolated cDNA copies of tubulin RNAs from the chicken, over 75 tubulin genes from at least twenty organisms have been sequenced. This body of data allows, with some reasonable confidence, generalizations to be made about tubulin gene structure and expression. The literature pertaining to this work has been reviewed extensively by others. The reviews by Raff[5], Cleveland & Sullivan[6], and Sullivan[7] are particularly thorough, and readers are referred to them and the vast literature cited within. What follows here is a distillation of some of the more provocative aspects of this work.

The genes for α- and β-tubulin occur, in all the organisms for which data exists, in small, heterogeneous gene families, with between one and six α- and one and six β-tubulin genes within each family. There is no evidence that the α- and β-tubulin genes are linked within genomes. Both α- and β-tubulin genes encode polypeptides of between 440 and 450 amino acid residues. The α-tubulins are very similar to β-tubulins (about 40% homology in those cases where comparisons have been made). It is likely that these genes diverged from a single ancestral tubulin gene. Furthermore, detailed studies of both vertebrate and invertebrate tubulin genes have shown that, within either of the α- or β-tubulin gene families, there is quite limited divergence of the products of tubulin genes, even in widely divergent species.

Tubulin genetics thus reveals a intriguing paradox: tubulin genes produce a highly conserved product, yet multiple tubulin genes are maintained in virtually all organisms. Why is this so? It could be that tubulin isotypes within an organism are functionally different, and these functions require different tubulins, or it could be that different tubulins are needed during different developmental stages. Data that has accumulated to date, particularly with respect to the β-tubulin gene family, suggests that, although many β-tubulins can be used interchangeably in cells (which would be expected if the divergence among tubulins is inconsequential to its function), there is good evidence that some tubulins are not functionally equivalent.

β-Tubulin gene expression and utilization of β-tubulin isotypes

In vertebrates, there are six recognizable classes of β-tubulins which can be identified by virtue of their amino acid sequences and by their patterns of expression[6,7]. These β-tubulins are very similar in sequence (greater than 90% homology in some cases) but they do share a variable region between amino acids 30 and 100, and uniformly have a hypervariable C-terminal region downstream from amino acid 430; this region is important for the binding of microtubule-associated proteins (MAPs), as will be discussed later. It is clear that the isoforms are not expressed uniformly within all the tissues of the organism. In chicken, for example, one isoform is found in abundance in many tissues, others are found only in neuronal tissues, and one divergent β-isoform is found only in haemopoietic tissue.

This pattern of developmentally regulated β-tubulin gene expression is found in other organisms as well. In some of these cases, the regulation seems to be quite tight. In the fruit fly, *Drosophila melanogaster*, the β2-tubulin isotype is found only in the germ line cells of the testis, whereas at least two of the remaining three β isotypes occur in many tissues. In the slime mold, *Physarum polycephalum*, the amoebal phase of the life cycle has only one electrophoretically distinct β isoform, whereas the plasmodium has an additional divergent β-tubulin isotype, which appears just as soon as the cell has become committed to the plasmodial developmental program. In other organisms, however, there is expression of tubulin gene products which is redundant. For example, in the nematode worm *Caenorhabditis elegans* one of the three β-tubulin genes, called *ben-1*, a gene which encodes a tubulin which is constitutively expressed, can be entirely deleted without affecting the growth, movement, or reproduction of the organism[8].

These examples seem to make the point that the pattern of tubulin gene expression is not a simple case of a particular gene product being used to build a particular subset of microtubules. The fact that some β-tubulins can be deleted entirely without effect seems to say that there is a degree of plasticity of tubulin isotype utilization; that is, that microtubules might simply be polymers of the tubulins that are available in the cell. In fact, heterologous β-tubulins that are microinjected into cells become incorporated into microtubules with the same efficiency as for native tubulins, and are, furthermore, often incorporated into *all* the microtubules in the cell. But a crucial recent experiment, in *Drosophila melanogaster*, provides convincing evidence that β-tubulins are *not* necessarily functionally equivalent[9]. As mentioned above, the β_2 isotype is found in the male germ cell line in flies, and is used to construct all the microtubule classes in these cells, including the cytoplasmic microtubules, the spindle, and the flagellar axoneme. The β_3 isotype is found in many tissues of the fly but *not* in the germ cells. If the β_3 gene is introduced into the germ cells under conditions wherein the normal β_2 is not expressed, only a small subset of cytoplasmic micro-tubules is produced. If a hybrid β_2–β_3 gene is expressed, the cytoplasmic and spindle microtubules are normal, but the development of the flagellum is impaired, even if the amount of β_3-tubulin is only 20% of the total tubulin pool. This result seems to indicate that, at least for the construction of flagellar microtubules, β-tubulins are *not* functionally equivalent.

Thus, the evidence with respect to β-tubulins suggests that there is no real rule of thumb that can be applied that will predict whether a cell can make catholic use of available β-tubulin isotypes, or whether there is a strict isotype requirement.

α-Tubulins and post-translational modifications

Another way to introduce biochemical specificity into a population of tubulin mole-cules is to chemically modify them, and thus "mark" them for some specific function or destination in cells. In fact, the α-tubulins are chemically modified, and thus "marked", by cytoplasmic enzyme-catalysed reactions which seem to modify α-tu-bulin specifically in either the free α–β heterodimer form, or in the assembled micro-tubule. Two such post-translational modifications have been described: these are acetylation/deacetylation and tyrosination/detyrosination (Figure 3).

The acetylation of α-tubulin was first explored in detail in the unicellular green alga, *Chlamydomonas reinhartdii*. This organism has two flagella which possess a distinct α-tubulin isoform (α_3). Rosenbaum and his colleagues[10] showed that the predominant cytoplasmic form of α-tubulin (α_1) was unacetylated, but that the α_3 isoform, which is the predominant (but not unique) form of α-tubulin in the flagella, contained an acetylated lysine at position 40 in the α-tubulin polypeptide. Interestingly, it was shown that the acetylation reaction (catalysed by a tubulin acetyltransferase) favours the *assembled* form of tubulin; thus the acetylation reaction occurs in the flagellar axoneme itself. The preponderance of the unacetylated form in the normal cytoplasmic tubulin pool, and under conditions wherein the flagella were induced to reabsorb, indicates that disassembly is coincident with the deacetylation reactions. Subsequent work has shown that acetylation of flagellar α-tubulin is widespread (and probably universal).

Several classes of non-flagellar microtubules contain acetylated α-tubulin. In general, these microtubules are those that are stable, and reasonably resistant to drug-induced disassembly. In *Trypanosoma*, for example, the entire sub-pellicular array of micro-tubules is uniformly acetylated, and in mammalian brain, acetylated microtubules have been shown in axons (although curiously not in the dendrites). It is also known that there are cells in which only deacetylated microtubules exist. In some cultured mammalian cells, and in the plasmodial stage of *Physarum*, there are no acetylated microtubules. This observation, taken together with the fact that some α-tubulin genes do not have the necessary lysine residue at position 40, indicate the acetylation of α-tubulin is not necessary for assembly of microtubules.

The process of tyrosination/detyrosination of α-tubulin is also most likely related to the age or stability of microtubules. Barra[11] originally discovered that tubulin can be tyrosinated by the action of a tubulin-tyrosine ligase (TTL), and can be detyrosinated by the action of a tubulin carboxypeptidase (TCP). The reaction may either add or remove a C-terminal tyrosine (or sometimes a phenylalanine). Many α-tubulin genes,

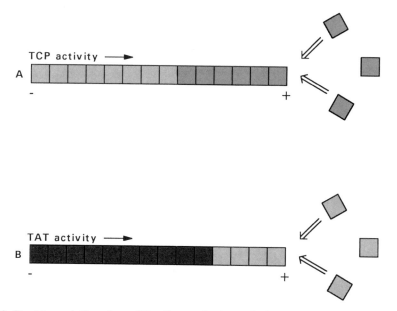

Figure 3. Post-translational modifications of microtubules
(A) Tyrosination/detyrosination. Tyrosinated dimers (shown as pink blocks) are added to the "+" end of the growing microtubule. These will have been tyrosinated by the action of tubulin tyrosine ligase (TTL). Tubulin carboxypeptidase (TCP) detyrosinates (grey blocks) the α-tubu-lins of the heterodimers once they are assembled into a microtubule. This activity proceeds from the "−" to the "+" end of the tubule. Thus tyrosination in a microtubule indicates that recent polymerization has occurred. (B) Acetylation/deacetylation occurs in a similar manner. De-acetylated tubulin heterodimers (grey blocks) are added to the growing end of the microtubule, and tubulin acetyltransferase (TAT) progressively acetylates them (red blocks) once they are assembled into the microtubule.

in fact, encode a C-terminal tyrosine, which could readily serve as a substrate for TCP. But, many α-tubulin genes do not encode a C-terminal tyrosine, so either the first reaction of these tubulins is to be tyrosinated, or they simply may not participate in the reactions at all.

TTL shows a marked preference for the unassembled dimer, whereas TCP prefers the assembled microtubule as a substrate. This asymmetry of tyrosination/detyrosination reactions may be important in the regulation of tubulin assembly. In fact, there is good reason to believe that tyrosination may be a useful "marker" for the age of a microtubule. Simply put, newly formed microtubules would be richer in the tyrosinated isoform, since detyrosination occurs sometime *after* assembly. If tissue culture cells are placed in the cold, or are treated with a microtubule depolymerizing drug such as colchicine, or nocodazole, the cellular microtubules reversibly depolymerize. Upon a return to normal temperatures, or upon the removal of the drug, the microtubules reassemble rapidly, *but* detyrosination does not occur until about 25 minutes *after* reassembly. The best example of this age-related tyrosination as it occurs *in vivo* comes from detailed study of the sub-pellicular and flagellar microtubules in *Trypanosoma brucei*[12]. Immunofluorescence images (using an antibody called YL1/2, which is specific for tyrosinated tubulin) showed two striking features of tyrosinated tubulin distribution: tyrosinated tubulin occurred only in the posterior one-third of the sub-pellicular microtubules (where new microtubules were being formed), and, as the single flagellum was replicated, the daughter flagellum was rich in tyrosinated tubulin. The staining disappeared with time, as the flagellum approached its "mature" length.

THE CELL BIOLOGY OF TUBULIN

As the first few sections of this review have described, microtubule research in the past 20 years has made significant progress towards understanding the biochemistry and genetics of tubulin. But what of the biology of tubulin? Microtubules are crucial in the intracellular distribution of cell organelles (such as mitochondria, endoplasmic reticulum, the Golgi and other membranous organelles), and appear to be involved, somehow, in maintaining cell shape. Just how are these microtubule-based processes achieved? The next sections describe some of the research on these processes, with a particular emphasis on the properties of some important microtubule-associated proteins.

Microtubule-associated proteins

As described earlier, assembly/disassembly cycling can be used to purify tubulin. But the product of this purification is not homogeneous. A number of proteins copurify together with microtubules. These proteins were originally termed "microtubule-associated proteins" (MAPs). The major MAPs have been named accordingly, and are called MAP1 (380 kDa), MAP2 (280 kDa), and the "tau" proteins (55–62 kDa). The general properties and characteristics of these proteins have been reviewed elsewhere[13]. Although most of these proteins have been known for the last 15 years, little is known about their function. The fact that they are particularly abundantly bound to microtubules isolated from brain tissue suggested that they may be important for microtubule stabilization, crosslinking of bundles of microtubules, and for connecting microtubules to other cell components. In fact, the actual experimental evidence for these guesses was missing until quite recently.

MAPs have different subcellular localizations amongst a set of cellular microtubules. For example, MAP2 is predominantly present in the dendrites, whereas tau is found in the axons of nerve cells. During cell development, different arrangements of MAPs are formed by differential expression and splicing of MAP gene products. In some cases, the MAPs expressed in earlier stages of development have a lower affinity for microtubules and their ability to crosslink them is also weaker as compared to later stages in development[14]. In addition, MAPs and the tau proteins appear to have qualitative effects on the microtubule lattice in cells. In a recent experiment, mammalian fibroblasts were transfected with cDNAs for MAP2 and tau proteins. Normally, MAP2 and tau proteins are *only* expressed in brain tissue. After being transfected, these cells showed an altered morphology and extensive arrays of crosslinked microtubules[15,16]. It can be concluded from these experiments that tau and MAP2 have crosslinking functions and are involved in maintaining a cell morphology which requires microtubule bundling.

There are many interactions between microtubules and various cell organelles that have been described in mostly anecdotal accounts. MAPs are often implicated in these accounts, but much work is left to be done in order to understand properly the function and role of MAPs in the cytoskeleton. It seems reasonable to predict that the subcellular distribution of MAPs allows the cell to develop different functional cytoskeletal domains (e.g. dendrites and axon). Regulation of MAP genes no doubt occurs, allowing the cell to produce MAPs with different properties for diverse structural and functional roles which may be required in tissue differentiation. The question which asks if particular MAPs associate with particular subsets of microtubules will likely have an interesting answer. Readers are referred to the recent review by Matus[17] for a comprehensive overview of MAPs and their role in cell development.

Molecular motors

There has long been a suspicion that the microtubules in cells can act as either a passive "guide track" for the movement of cellular organelles, or could actually "drive" the organelles themselves via some kind of microtubule-based motor. This suspicion grew in stature with the discovery and extensive biochemical description of *dynein*, a microtubule-associated ATPase, which provides the force for the flagellar beat (as described in full in the volume by Warner *et al.*[18]). An additional group of putative microtubule-associated motor proteins was soon discovered to exist elsewhere in the cell. For example, there are proteins which are thought to move organelles along microtubules, to extend the shape of the endoplasmic reticulum, and to move chromosomes along the spindle. These force-generating proteins are thus localized in the cytoplasm, in spindle poles, on vesicles and on kinetochores, in addition to the flagella.

Two of these proteins (*kinesin* and *MAP1C*) were shown to move membranous organelles *in vitro* along microtubules; in fact, if coverslips are covered in a layer of kinesin or MAP1C, an entire microtubule will slide over them! A third protein, called *dynamin*, supports the sliding of microtubules along each other. MAP1C, which is closely related to the flagellar dynein (and therefore often called cytoplasmic dynein),

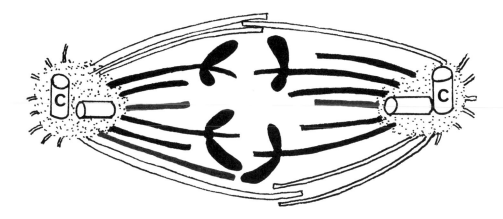

Figure 4. A mitotic spindle at anaphase
Pairs of centrioles (C) lie at either pole of the spindle, and microtubules radiate out from them.
Two types of microtubules can be identified: the polar microtubules (shown as white with black
outlines), and the kinetochore microtubules (shown in black or red). The polar microtubules
extend from the spindle poles to the equatorial plane of the spindle. The kinetochore microtubules extend from the poles toward the chromosome kinetochores (the ones shown in black).
Some kinetochore microtubules are shown growing outwards from the poles, and two in each
half-spindle have made contact with chromosomal kinetochores. The microtubules shown in
red are rapidly depolymerizing, and are thus getting shorter (dynamic instability). The chromosomes are depicted as having moved a short distance toward the poles along each of their
kinetochore microtubules.

produces movement of vesicles towards the minus end of microtubules (so-called
retrograde transport), while kinesin produces the movement of vesicles towards the
plus end of microtubules (*anterograde transport*). Initially kinesin and MAP1C were
isolated from brain tissue, but it is now clear that these proteins are more widespread
in occurrence[19]. What is less clear is the molecular mechanism of movement, how
these motors are regulated, and if these are the only motors present in cells. In fact,
some organisms possess motor proteins which are similar to the ones so far described,
but which can drive movement in *either* direction along a microtubule[20,21]. It will be
interesting to learn, as the story of these proteins unfolds, whether these motor proteins
support transport, in whichever direction, along *all* microtubules, or along only a
particular subset of microtubules.

Mitosis

There is perhaps no more familiar process in biology than mitosis, the events which
ensure that chromosomes are equally distributed between the two resulting daughter
cells. Any student of biology will know that the chromosomes attach to the *mitotic
spindle*, a highly ordered set of microtubules which is assembled specifically for this
purpose, and that the chromosomes then somehow migrate along this spindle to the
poles of the dividing cell. But the very structure of the spindle, packed with microtubules of apparently various lengths, and the chromosomes themselves, have made
resolution of the details of chromosome attachment and movement difficult.

It is clear that there are two types of microtubules in the spindle: the *polar microtubules* which run from the spindle poles toward the metaphase plane, and the *kinetochore microtubules,* which connect the spindle poles to the kinetochores of the chromosomes (Figure 4). Recent work[22] has shown that microtubules grow from the spindle poles in direction of the metaphase plate and once there get captured by kinetochores before any chromosome movement occurs. This observation answers the question about the direction of microtubule growth in the spindle and provides an explanation of how the contacts between spindle poles and kinetochores are established. As the principle of dynamic instability predicts, the kinetochore microtubules continually grow and shrink, which enables them to probe throughout the cytoplasm until they attach to a kinetochore[23].

It is still an open question whether the poleward movement of chromosomes is mediated by motor proteins found on the kinetochores[24], or by microtubule depolymerization[25]. The orientation of the chromosomes in metaphase is still to be resolved. Recent work suggests that the chromosome alignment at the metaphase plate is the net result of pulling forces of the chromosomes and pushing forces of the kinetochore microtubules[26].

There are many other detailed and important questions about mitosis yet to be addressed, none of which can be described in this short review. A detailed view of the organization and behaviour of microtubules in the mitotic spindle is given in the volume by Warner *et al.*[27].

Microtubule organizing centres (MTOCs)

What to do with the ends of microtubules is more than just a trivial question. Given that microtubules can assemble and disassemble rapidly, and that some have distinct polarity, it is not surprising that *microtubule organizing centres (MTOCs)* are found in most eukaryotic cells[28]. Nearly all cells have a *centrosome* near the nucleus. In many cells, the centrosome is occupied by a pair of centrioles, which in turn is surrounded by poorly defined dense fibres and other material. In dividing cells, the centrosome splits into two spindle poles, which subsequently organize the complex microtubule array of the spindle. The *basal body* lies at the base of cilia and flagella in those organisms that have them. In organisms which lack ciliated or flagellated cells, basal bodies and centrioles are absent, but the centrosome (or its equivalent) still exists.

The most obvious function of an MTOC is to anchor (and presumably stabilize) the minus end of a microtubule. Whether it is the centrioles themselves, or the electron dense material, called "pericentriolar material", which has this role is not known (although, of course, cells which lack centrioles can do just fine without them!). It is clear that the microtubules near to the centrosomes are stable to cold and drug depolymerization, and, as mentioned earlier, the γ-tubulin isoform is found located near these centres. In addition, the MTOCs may possibly have a role in the spatial organization of microtubules throughout the cell cycle. MTOCs contain a variety of different proteins which appear to vary in number, which can be post-translationally modified, or phosphorylated[29]. There is no doubt that research in the next few years will clarify the role of MTOCs, and will do much to help understand how microtubules are organized in cells.

FUTURE DIRECTIONS

Many tubulins and MAPs have been described at the polypeptide and gene level, and some three-dimensional molecular structure has been deduced at the electron microscope level. It is likely that the future of microtubule research is to look at the interactions between microtubules, MAPs and cell organelles. It is our view that the modern techniques of molecular biology, biochemistry, and video enhanced microscopy of living cells will rapidly make inroads into these problems of spatial and dynamic microtubule organization and regulation.

Bodo Lange is supported by a scholarship of the Boehringer Ingelheim Fonds and the SERC. Work in the Manchester laboratory is funded by the SERC. This investigation received financial support from the UNDP/World Bank/WHO Special Programme for Research and Training in Tropical Diseases.

REFERENCES

1. Mandelkow, E. & Mandelkow, E.-M. (1989) in *Cell Movement*, vol. 2, pp. 23–45, Alan R. Liss Inc., New York

2. Oakley, C.E. & Oakley, B.R. (1989) Identification of γ-tubulin, a new member of the tubulin superfamily encoded by *mipA* gene of *Aspergillus nidulans. Nature (London)* **338**, 662–664

3. Oakley, B.R., Oakley, E.C., Yoon, Y. & Jung, M.K. (1990) γ-Tubulin is a component of the spindle pole body that is essential for microtubule function in *Aspergillus nidulans. Cell* **61**, 1289–1301

4. Kirschner, M.W. & Mitchison, T.J. (1986) Beyond self-assembly: from microtubules to morphogenesis. *Cell* **45**, 329–342

5. Raff, E.C. (1984) Genetics of microtubule systems. *J. Cell Biol.* **99**, 1–10

6. Cleveland, D. W. & Sullivan, K.F. (1985) Molecular biology and genetics of tubulin. *Annu. Rev. Biochem.* **54**, 331–365

7. Sullivan, K.F. (1988) Structure and utilization of tubulin isotypes. *Annu. Rev. Cell Biol.* **4**, 687–716

8. Driscoll, M., Dean, E., Reilly, E., Bergholz, E. & Chalfie, M. (1989) Genetic and molecular analysis of a *Caenorhabditis elegans* β-tubulin that conveys benzimidazole sensitivity. *J. Cell Biol.* **109**, 2993–3003

9. Hoyle, H. & Raff, E.C. (1990) Two *Drosophila* β-tubulin isoforms are not functionally equivalent. *J. Cell Biol.* **111**, 1009–1026

10. Greer, K. & Rosenbaum, J.L. (1989) in *Cell Movement,* vol. 2, pp. 47–66, Alan R. Liss Inc., New York

11. Barra,H.S., Rodriguez, J.A., Arce, C.A. & Caputto, R. (1973) A soluble preparation from rat brain that incorporates into its own proteins ^{14}C-arginine by a ribonuclease-sensitive system and ^{14}C-tyrosine by a ribonuclease-insensitive system. *J. Neurochem.* **20**, 97–108

12. Sherwin, T., Schneider, A., Sasse, R., Seebeck, T. & Gull, K. (1987) Distinct localization and cell cycle dependence of COOH terminally tyrosinolated α-tubulin in the microtubules of *Trypanosoma brucei brucei. J. Cell Biol.* **104**, 439–446

13. Olmsted, J.B. (1988) Microtubule-associated proteins. *Annu. Rev. Cell Biol.* **2**, 421–457

14. Himmler, A. (1989) Structure of the bovine tau gene: alternatively spliced transcripts generate a protein family. *Mol. Cell Biol.* **9**, 1389–1396

15. Kanai, Y., Takemura, R., Oshima, T., Mori, H., Ihara, Y., Yanagisawa, M., Masaki, T. & Hirokawa, N. (1989) Expression of multiple tau isoforms and microtubule bundle formation in fibroblasts transfected with tau cDNA. *J. Cell Biol.* **109**, 1173–1184

16. Lewis, S.A., Ivanov, I.E., Lee, G.H. & Cowan, N.J. (1989) Microtubule organization in dendrites and axons is determined by a short hydrophobic zipper in microtubule-associated protein MAP2 and tau. *Nature (London)* **342**, 498–505

17. Matus, A. (1988) Microtubule-associated proteins. *Annu. Rev. Neurosci.* **11**, 29–44

18. Vale, R.D. (1990) Microtubule-based motor proteins. *Curr. Opinions Cell Biol.* **2**, 15–22

19. Warner, F.D., Satir, P. & Gibbons, I.R. (eds.) (1989) *Cell Movement: Vol.1, The Dynein ATPases*, Alan R. Liss, Inc., New York.

20. Malik, F. & Vale, R. (1990) A new direction for kinesin. *Nature (London)* **347**, 713–714

21. McIntosh J.R. & Porter, M.E. (1989) Enzymes for microtubule-dependent motility. *J. Biol. Chem.* **264**, 6001–6004

22. Hayden, J.H., Bowser, S.S. & Rieder, C.L. (1990) Kinetochore capture astral microtubules during chromosome attachment to the mitotic spindle: direct visualization in live newt lung cells. *J. Cell Biol.* **111**, 1039–1045

23. Mitchison, T.J. (1990) The kinetochore in captivity. *Nature (London)* **348**, 14–15

24. Hyman, A.A. & Mitchison, T.J. (1991) Two different microtubule-based motor activities with opposite polarities in kinetochores. *Nature (London)* **351**, 206–211

25. Koshland D.E., Mitchison, T.J. & Kirschner, M.W. (1988) Polewards chromosome movement driven by microtubule depolymerisation *in vitro*. *Nature (London)* **331**, 499–504

26. Salmon E.D. (1989) in *Cell Movement*, vol. 2, pp. 431–440, Alan R. Liss Inc., New York

27. Warner, F.D. & McIntosh, J.R. (eds.) (1989) *Cell Movement*, vol. 2, chapters 23–28, Alan R. Liss Inc., New York

28. Huang, B. (1990) Genetics and biochemistry of centrosomes and spindle poles. *Curr. Opinions Cell Biol.* **2**, 28–32

29. Tousson, A., Zeng, C., Brinkley, B.R. & Valdivia, M.M. (1991) Centrophilin, a novel mitotic spindle protein involved in microtubule nucleation. *J. Cell Biol.* **112**, 427–440

3

Plant signal perception and transduction: the role of the phosphoinositide system

Bjørn K. Drøbak

Department of Cell Biology, John Innes Institute, John Innes Centre for Plant Science Research, Colney Lane, Norwich NR4 7UH, U.K.

INTRODUCTION

In common with all other living organisms, plants have to cope with a hotch-potch world and are constantly fighting the losing battle against the second law of thermodynamics. Prolonging the period till final surrender is often of vital importance for the continuation and proliferation of the species. Unlike many of their eukaryotic colleagues, plants are generally unable to escape from a given location when conditions change and become unfavourable. Thus there is particular need for plants to be able to register and adapt to a changing environment.

Cellular mechanisms for receiving, and interpreting, information from the surrounding environment were developed during the early stages of evolution. The perception of change often has the purpose of inducing a cellular response. Since a firm barrier needs to be maintained towards the outside world, and as closely controlled intracellular physicochemical conditions are important for metabolic processes, it could prove both hazardous, and indeed difficult, for a cell to allow external messages direct access to its interior. As a consequence, a large number of cell–environment interactions have evolved in which signals are registered first at the exterior surface of the plasma membrane. In order for the signal to be conveyed to the appropriate response elements inside the cell a sequence of events — often referred to as signal

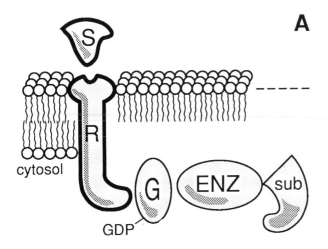

A

Figure 1. Outline of receptor-coupled transmembrane signalling processes common to several eukaryotic signal transducing systems
Panel A, unstimulated cell; panel B, stimulated cell. Symbols: S, signal; R, receptor molecule; G, regulatory GTP-binding protein; ENZ, enzyme; sub, substrate.

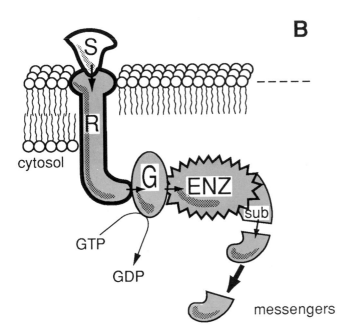

B

transduction — needs to occur. A number of different eukaryotic signal transduction pathways are currently known and it has recently become apparent that several of the most prominent pathways have a number of biochemical features in common. A schematic representation of such systems is given in Figure 1.

SIGNAL TRANSDUCTION

The initial step in the cell's recognition of the arrival of a signal at the cell surface is achieved by a molecular interaction between the signal (S) and an appropriate membrane-associated receptor protein (R). This interaction may in many cases convert one or more membrane associated GTP-binding proteins (G-proteins) into their active

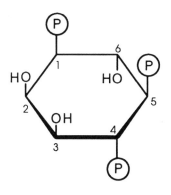

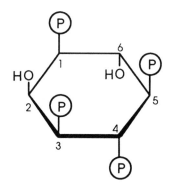

D-*myo*-inositol 1,4,5-trisphosphate   **D-*myo*-inositol 1,3,4,5-tetrakisphosphate**

form(s). The interaction between the signal–receptor complex and the GTP-binding protein(s) leads to the activation of an enzyme(s) localized on the cytosolic face of the plasma membrane. As a result of this enzyme activation, molecules are formed which are capable of carrying the message conveyed by the extracellular signal to intracellular response-elements. Although these message-carrying molecules or ions, are produced relatively late in the transduction cascade, they are often referred to as "second messengers". Several second messengers are now known to be functioning in mammalian cells. These include calcium ions, cyclic AMP, cyclic GMP, 1,2-diacyl-glycerol (DG), inositol 1,4,5-trisphosphate [Ins(1,4,5)P_3] and probably inositol 1,3,4,5-tetrakisphosphate [Ins(1,3,4,5)P_4]. Of these only the Ca^{2+} ion is known, beyond any doubt, to be involved in transmembrane signalling in plants. However, in the last few years evidence has been rapidly accumulating suggesting that Ins(1,4,5)P_3 also plays a role as a second messenger in plant cells.

CALCIUM AS A SECOND MESSENGER IN PLANTS

The divalent nature of Ca^{2+} and the lack of stringent crystal field requirements for Ca^{2+} chelate formation allows this ion to interact with a wide range of biological molecules with very fast association/dissociation kinetics. The ability of Ca^{2+} or Ca^{2+}–Ca^{2+}-binding protein complexes (e.g. Ca^{2+}–calmodulin) to modulate cellular processes has resulted in the development of signalling systems where Ca^{2+} acts as a second messenger.

The Ca^{2+} concentrations surrounding most plant (and mammalian) cells are in the low millimolar region. The concentration (or more precisely, activity) of Ca^{2+} in the cytosol needs to be kept very low in "unstimulated" cells to avoid constant activation of Ca^{2+}-dependent response elements. Studies using intracellular Ca^{2+}-sensitive dyes and microelectrodes have shown that the Ca^{2+} activity in the cytosol of both mammalian and plant cells is maintained at around 50–200 nM when the cell is not being stimulated[1]. The very steep electrochemical gradient across the plasma membrane, and the fact that electrical driving forces for cytosolic Ca^{2+} entry also exist, necessitate the presence of ion transporting systems with the ability of rapidly removing Ca^{2+} from the cytosol. The Ca^{2+} transporting systems in plant cells believed to be of most importance for control of cytosolic Ca^{2+} at low activities are the plasma membrane and endoplasmic reticulum Ca^{2+}-ATPases and the tonoplast H^+/Ca^{2+}-antiport. It was previously thought that mitochondria played a role in the regulation of cytoplasmic Ca^{2+}. However, it is now generally accepted that, although these organelles possess Ca^{2+} transport systems, their characteristics are such that they are unlikely to play any major role in the control of Ca^{2+} levels in the cytoplasm of unstimulated cells.

Upon stimulation of cells by a wide range of extracellular signals a rapid increase in cytosolic Ca^{2+} ensues. Since Ca^{2+} can neither be synthesized nor broken down the only mechanisms by which a change in Ca^{2+} activity can be achieved is by altering the cellular Ca^{2+}-buffering capacity or by movement of Ca^{2+} from one compartment to another.

The debate about the source of Ca^{2+} mobilized during signalling events has been going on for a considerable length of time amongst researchers and is still not completely resolved. In the early 1980s the general view was that Ca^{2+} was somehow introduced into the cytoplasm from the extracellular medium via signal-sensitive Ca^{2+}-channels located in the plasma membrane. However, the discovery around 1983 of the phosphoinositide signalling system by Berridge, Irvine, Michell and their co-workers has dramatically changed this view[2].

THE PHOSPHOINOSITIDE SYSTEM

Upon stimulation of mammalian cells by selected agonists known to induce a rise in intracellular Ca^{2+}, it was found that a minor inositol-containing phospholipid,

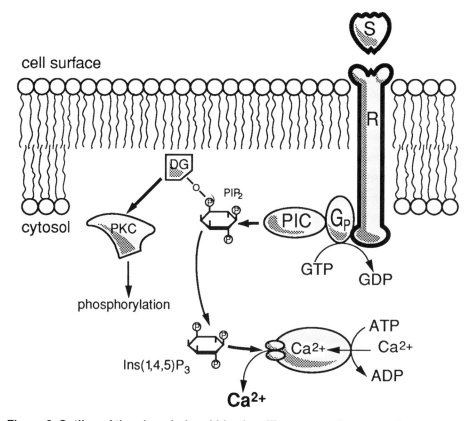

Figure 2. Outline of the phosphoinositide signalling system in mammalian cells

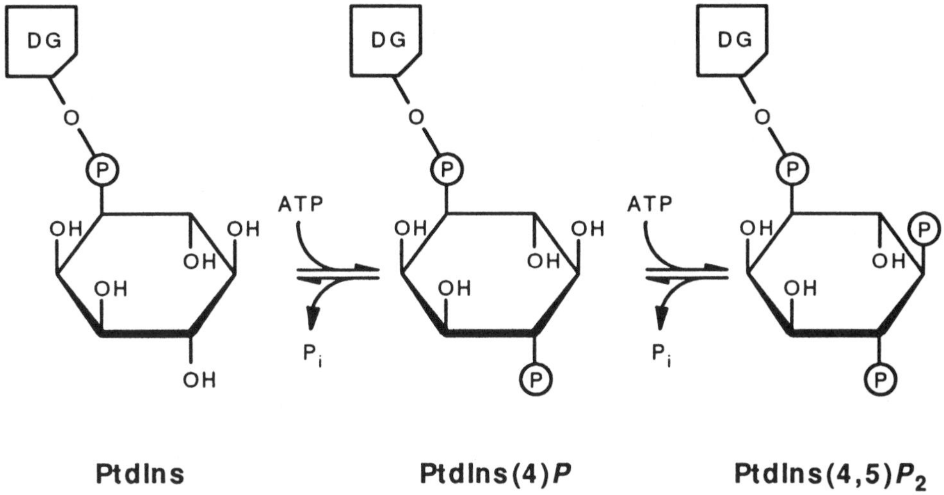

Figure 3. Interconversion between phosphatidylinositol and polyphosphoinositides

phosphatidylinositol 4,5-bisphosphate [PtdIns(4,5)P_2], localized in the inner leaflet of the plasma membrane, is hydrolysed by the enzyme phosphoinositidase C (phospholipase C). This results in the formation of two molecules, diacylglycerol (DG) and Ins(1,4,5)P_3. DG remains in the plasma membrane matrix where it activates a protein kinase (protein kinase C). Ins(1,4,5)P_3, which is highly polar, diffuses from the plasma membrane to intracellular membrane-bounded Ca^{2+} stores (likely to be associated with the endoplasmic reticulum) where it interacts with specific Ins(1,4,5)P_3 receptor molecules. The Ins(1,4,5)P_3–receptor interaction results in the opening of Ca^{2+} channels and Ca^{2+} is released into the cytosol. Thus, certain agonists are able to trigger a bifurcated cellular response involving both activation of protein kinase C and a rise in cytosolic Ca^{2+}. The use of two transduction strands opens the possibility for cross-talk and results in a high degree of flexibility in cellular interpretation. This may explain why the phosphoinositide signalling system is involved in the transduction of a large number of signals in a wide range of eukaryotes. The phosphoinositide signalling system fits the general scheme outlined in Figure 1. A more detailed description of this system is given in Figure 2.

PtdIns(4,5)P_2 is formed by a two-step phosphorylation of phosphatidylinositol (PtdIns) (Figure 3). PtdIns is first phosphorylated in the 4 position of the inositol ring resulting in the formation of phosphatidylinositol 4-phosphate [PtdIns(4)P]. This lipid is then further phosphorylated in the 5 position yielding PtdIns(4,5)P_2. Two phosphohydrolases work concomitantly with the 4- and 5-hydroxykinases so PtdIns(4)P and PtdIns(4,5)P_2 are constantly being formed and degraded. This has been termed the "futile" phosphoinositide cycle but it is now clear that this cycle is likely to be anything but futile. Both DG and Ins(1,4,5)P_3 are rapidly hydrolysed, so removal of the signal causing the activation of phosphoinositidase C results in deactivation of protein kinase C and a return of cytosolic Ca^{2+} concentrations to unstimulated levels, courtesy of the Ca^{2+} transport systems.

Since Ca^{2+} also acts as a second messenger in plant cells it is an obvious possibility that fluxes are controlled by a system corresponding to the mammalian phospho-inositide signalling system. The first direct evidence supporting this idea came from experiments investigating the effects of $Ins(1,4,5)P_3$ on Ca^{2+} fluxes across isolated plant membranes. It was found that $Ins(1,4,5)P_3$ in molar concentrations is able to cause rapid release of Ca^{2+} from plant microsomal vesicles[3,4]. Assuming that these vesicles primarily are right-side out and derived from organelles, the Ca^{2+} release is equivalent to emptying of intracellular Ca^{2+} stores into the cytosol. These findings gave the impetus to plant scientists to start a search for both structural and functional aspects of a phosphoinositide signalling system in plant cells. Research carried out in the last 5 or 6 years has shown that many components of a mammalian-type phosphoinositide signalling system are present in plant cells (see Table 1). It has been suggested that the phosphoinositide signalling system may function in several im-portant transduction events in plant cells, and evidence is accumulating that supports this view[11–15].

Now that it is recognized that many apparent similarities exist between the mam-malian and the plant phosphoinositide signalling system the time is ripe to address the question: "What are the differences?". Three of the most prominent areas of dif-ference are discussed in the following sections.

POLYPHOSPHOINOSITIDE METABOLISM

Several types of experiments have been carried out to investigate the metabolism of PtdIns$P$ and PtdInsP_2 in plant cells. Experiments *in vitro* using isolated membrane fractions as enzyme source and $[\gamma\text{-}^{32}P]$ATP as phosphate donor have demonstrated that both PtdIns 4-hydroxykinase and PtdIns(4)P 5-hydroxykinase are present and can utilize both endogenous and exogenous substrates[7]. Such studies have identified the plasma membrane as one of the main sources of polyphosphoinositide kinases, and it has been observed that the PtdIns kinase is around 10-fold more active than the PtdIns$P$ kinase under optimum assay conditions. In this regard the plant poly-phosphoinositide kinases are not grossly dissimilar to the mammalian enzymes. A quite different picture however emerges when *in vivo* labelling studies are carried out. When cultured plant cells or tissues are incubated with $[^{32}P]$orthophosphate it is found, after short incubation times, that a large percentage of total phospholipid label (approx. 30 %) is present in the monoester phosphate of PtdIns(4)P, indicating that this lipid is metabolized very rapidly (i.e. the PtdIns 4-hydroxykinase is highly

---

Table 1. Identified components of the plant phosphoinositide system

1. Ability of $Ins(1,4,5)P_3$ to release Ca^{2+} from intracellular stores[3,4]
2. Presence of phosphatidylinositol, phosphatidylinositol 4-phosphate and phosphatidylinositol 4,5-bisphosphate in plant membranes[5,6]
3. Presence of kinases/phosphatases involved in polyphosphoinositide turnover[5,7]
4. Presence of phospholipase(s) C capable of hydrolysing phosphatidylinositol 4,5-bisphosphate[8]
5. Presence of enzymes which in several respects resemble mammalian protein kinase C[5]
6. Presence of enzymes capable of rapidly metabolizing $Ins(1,4,5)P_3$[9,10]

active). In contrast PtdInsP_2 only incorporates very small amounts of label after both short- and long-term incubations[5]. The incorporation of label into PtdInsP_2 is far lower than one would expect on the basis of *in vitro* studies of PtdIns$P$ kinase activity. Several explanations for this apparent anomaly can be envisaged. The most likely explanation is that the chemical levels of PtdInsP_2 are extremely low. This would mean that however rapidly label is incorporated into PtdInsP_2 the total amount of incorporated radioactivity will always appear very minor when compared to the label found in the total phospholipid pool (except after extremely short labelling times! — but total label in the phospholipid pool at such times is so low that the investigation of this possibility is very difficult).

Some researchers see the very low levels (or slow rate of metabolism) of PtdInsP_2 as an obstacle to Ins(1,4,5)P_3/DG formation. This however is not necessarily the case as only very small amounts of PtdInsP_2 are needed to fulfil the role as second messenger precursor. It should be remembered that although PtdInsP_2 is a minor component of the total cellular phospholipid pool it is likely to be unequally distributed within the cell. If for instance all, or a major part, of cellular PtdInsP_2 is localized in the plasma membrane (which only accounts for 2–5 % of cellular membranes) while PtdIns and PtdIns$P$ are more equally distributed throughout the cell, the PtdIns/PtdIns$P$/PtdInsP_2 ratios in the plasma membrane could well approach "mammalian" levels.

Another possibility which needs considering is that the PtdIns$P$ kinase may be under direct control of the signal–response coupling complex and may only be active when an agonist interacts with its receptor. If this is the case, large quantities of Ins(1,4,5)P_3/DG could be produced in a short period of time without necessarily changing the chemical levels of PtdInsP_2. Simultaneous activation of phosphoinositidase C and the PtdIns$P$ kinase would result in a very rapid flux through the PtdInsP_2 pool. The rate of flux is often the important feature in signalling events, not absolute chemical amounts.

Two additional aspects of polyphosphoinositide metabolism and function deserve brief mentioning in this context. The traditional view of polyphosphoinositide formation/degradation involving plasma membrane associated kinases/phosphatases has recently been challenged. PtdIns in both plant and mammalian cells is synthesized intracellularly and then transported to the plasma membrane where it is incorporated into the bilayer. It has been found that PtdIns$P$ in rat hepatocytes *also* is synthesized intracellularly and transported to the plasma membrane. It is thought that the further phosphorylation to PtdInsP_2 occurs exclusively in the plasma membrane. Should a similar system be operating in plant cells several new hypotheses explaining rapid PtdIns$P$ turnover and low levels of PtdInsP_2 can be envisaged.

Considering the high rate of PtdIns$P$ turnover and the considerable amount of adenylate energy expended by the PtdIns 4-kinase in plant cells the question must be asked : "Is the function of PtdIns$P$ solely to act as precursor for PtdInsP_2 formation?". Some evidence suggesting that the answer to this question may be "no" has recently been obtained. It has been shown that polyphosphoinositides are capable of modulating the activity of the plasma membrane H$^+$-ATPase in sunflower hypocotyls[16] and several additional functions for polyphosphoinositides in plant cells are currently under investigation.

METABOLISM OF Ins(1,4,5)P_3

Sixtythree different isomers of D-*myo*-inositol phosphates are theoretically possible, and well over a dozen have been found to be present in different mammalian cells. This array of inositol phosphates is currently causing some problems for researchers in the mammalian field. Plant scientists are faced with the fact that inositol phosphate metabolism in plants is likely to have an even higher degree of complexity. In mammalian cells all known inositol phosphates are derived as a result of phosphoinositide hydrolysis. This imposes at least a degree of constraint on the InsP_x isomers produced. In plant cells at least one additional pathway for inositol phosphate production exists. This is the phytic acid (inositol hexakisphosphate) pathway. In plants significant amounts of phytic acid accumulate in seeds and other tissues and several important physiological roles for phytic acid are known. Apart from acting as a donor of orthophosphate during germination, phytic acid is also able to increase seed viability due to its strong antioxidative properties. At present little is known about the route of phytic acid formation and degradation in plants, except that D-*myo*-inositol 3-monophosphate probably is the precursor and that enzymes exist which are capable of stepwise phosphorylation of inositol phosphates and dephosphorylation of phytic acid[17]. It has recently been demonstrated that phytic acid in the slime mould *Dictyostelium* is formed by stepwise phosphorylation of *myo*-inositol and that this pathway bypasses the inositol phosphate isomers known to be involved in signal transduction[18]. Whether a similar biosynthetic route is followed in plant cells is currently not known.

In recent experiments the possible metabolic pathways of Ins(1,4,5)P_3 in plant cells have been investigated using cell-free extracts as enzyme sources. It has been shown that Ins(1,4,5)P_3 is rapidly metabolized into both higher and lower inositol phosphates. Whereas the main dephosphorylation product of Ins(1,4,5)P_3 in mammalian cells is Ins(1,4)P_2 it was found that the prevalent isomer produced by the plant enzymes is Ins(4,5)P_2. Although Ins(1,4)P_2 is also a dephosphorylation product of Ins(1,4,5)P_3 in plants, the Ins(1,4,5)P_3 5-phosphatase appears to be much less active than the 1-phosphatase[9]. In addition to these InsP_2s nine other metabolites were formed, including inositol and inositol tetrakisphosphate. These studies have made it clear that enzyme systems are present in plants which are capable of rapidly metabolizing Ins(1,4,5)P_3 using pathways not encountered in other eukaryotes. An overview of the possible pathways of inositol phosphate formation and degradation in plant cells is given in Figure 4.

At least 15 inositol phosphates of unknown isomeric configuration are present in many plant cells. Several of these have chromatographic properties very similar to the "messenger" inositol phosphate(s) and their hydrolysis products. This "background" of cellular inositol phosphates makes it very difficult to study rapid transients of message-carrying inositol phosphates in plant cells.

WHAT IS THE NATURE OF THE Ins(1,4,5)P_3-SENSITIVE Ca^{2+} POOL ?

Current evidence from studies using patch-clamp and indicator dye techniques strongly points towards the vacuole as being the Ins(1,4,5)P_3-sensitive Ca^{2+} store in plant cells[19]. Much more information is however needed before the vacuole can be accepted as the sole target for Ins(1,4,5)P_3-induced Ca^{2+} release in plant cells. Investigations of the spatial distribution of PtdIns(4,5)P_2, phosphoinositidase C, Ins(1,4,5)P_3

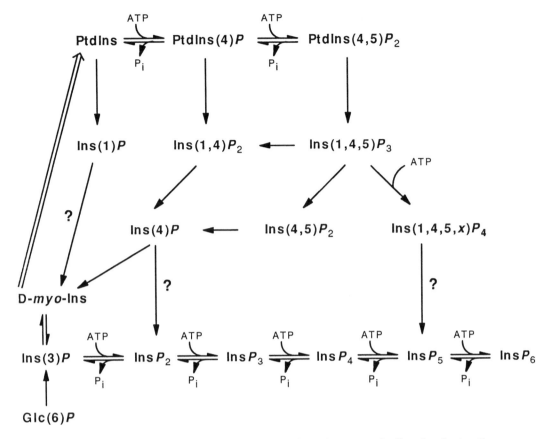

Figure 4. Likely pathways for inositol/inositol phosphate metabolism in plant cells
Arrows indicate enzymic steps which have been characterized *in vivo* or *in vitro*. All isomers
are numbered using the D-configuration. Arrows marked with "?" indicate potential pathways
which have not yet been firmly established. The diagram has been constructed on the basis of
data from references 5,6,7,8,9,10,13 and 17.

tonoplast receptors and Ins(1,4,5)P_3-metabolizing enzymes within the cell are clearly
necessary before a detailed assessment of temporal aspects of signal-induced Ca^{2+}
release is possible. Although the distance between the plasma membrane and the
tonoplast varies from cell type to cell type, and also within a cell, the average distance
is often small (100–200 nm). Ins(1,4,5)P_3 could thus easily diffuse from plasma membrane to tonoplast in a short period of time. The vacuole occupies approx. 90% of
the intracellular volume in many plant cells and the concentration of free Ca^{2+} is
typically 1–2 mM[1]. The enormous difference in both Ca^{2+} concentration and total Ca^{2+}
between the vacuole and the cytoplasm necessitates extremely meticulous cellular
control over the opening and closure of Ca^{2+} channels in the tonoplast. The failure
to maintain very stringent control of Ca^{2+} efflux will rapidly result in the build up
of lethal Ca^{2+} concentrations in the cytoplasm.

It is an open question why plant cells should have developed mechanisms for
signal transduction which in many respects resemble those of mammalian cells and
yet use a signal-sensitive Ca^{2+} pool which is so different from that found in most

other eukaryotes. As we learn more about the details of Ca^{2+} handling in plant cells it is likely that a much more complex picture with many nuances will emerge. That this is already happening is perhaps best illustrated by a very recent study of abscisic acid-induced Ca^{2+} fluxes in guard cells which points to the possibility of several signal-sensitive Ca^{2+} pools operating in concert[20].

REFERENCES

1. Felle, H. (1989) Ca^{2+}-selective microelectrodes and their application to plant cells and tissues. *Plant Physiol.* **91**, 1239–1242

2. Berridge, M.J. & Irvine, R.F. (1989) Inositol phosphates and cell signalling. *Nature (London)* **341**, 197–205

3. Drøbak, B.K. & Ferguson, I.B. (1985) Release of Ca^{2+} from plant hypocotyl microsomes by inositol(1,4,5)trisphosphate. *Biochem. Biophys. Res. Commun.* **130,** 1241–1246

4. Schumaker, K.S. & Sze, H. (1987) Inositol(1,4,5)trisphosphate releases Ca^{2+} from vacuolar membrane vesicles of oat roots. *J. Biol. Chem.* **262**, 3944–3946

5. Drøbak, B.K., Ferguson, I.B., Dawson, A.P. & Irvine R.F.(1988) Inositol containing lipids in suspension cultured plant cells. *Plant Physiol.* **87**, 217–222

6. Irvine, R.F., Letcher, A.J., Lander, D.J., Drøbak, B.K., Dawson, A.P. & Musgrave, A. (1989) Phosphatidylinositol(4,5)bisphosphate and phosphatidylinositol(4)phosphate in plant tissues. *Plant Physiol.* **89**, 888–892

7. Sommarin, M. & Sandelius, A.S. (1988) Phosphatidylinositol and phosphatidylinositol-phosphate kinases in plant plasma membranes. *Biochim. Biophys. Acta* **958,** 268–278

8. McMurray, W.C. & Irvine, R.F. (1988) Phosphatidylinositol(4,5)bisphosphate phosphodiesterase in higher plants. *Biochem. J.* **249**, 877–881

9. Drøbak, B.K., Watkins, P.A.C., Chattaway, J.A., Roberts K. & Dawson, A.P. (1991) Metabolism of inositol(1,4,5)trisphosphate by a soluble enzyme fraction from pea (*Pisum sativum*) roots. *Plant Physiol.* **95**, 412–417

10. Joseph, S.K., Esch, T. & Bonner, W.D. (1989) Hydrolysis of inositol phosphates by plant cell extracts. *Biochem J.* **264**, 851–856

11. McAinsh, M.R., Brownlee, C. & Hetherington, A.M. (1990) Abscisic induced elevation of guard cell cytosolic Ca^{2+} precedes stomatal closure. *Nature (London)* **343**, 186–189

12. Ettlinger, C. & Lehle, L. (1988) Auxin induces rapid changes in phosphatidylinositol metabolites. *Nature (London)* **331**, 176–178

13. Morse, M.J., Crain R.C. & Satter, R.L. (1987) Light-stimulated inositolphospholipid turnover in *Samanea saman* leaf pulvini. *Proc. Natl. Acad. Sci. U.S.A.* **84,** 7075–7078

14. Gilroy, S., Read, N.D. & Trewavas, A.J. (1990) Elevation of cytoplasmic calcium by caged calcium or caged inositol trisphosphate initiates stomatal closure. *Nature (London)* **346**, 769–771

15. Blatt, M.R., Thiel, G. & Trentham, D.R.(1990) Reversible inactivation of K^+ channels of *Vicia* stomatal guard cells following the photolysis of caged inositol(1,4,5)trisphosphate. *Nature (London)* **346**, 766–769

16. Memon, A.R. & Boss, W.F. (1990) Rapid light-induced changes in phosphoinositide kinases and H^+-ATPase in plasma membrane of sunflower hypocotyls. *J.Biol.Chem.* **265,** 14817–14821

17. Loewus, F.A. & Loewus, M.W. (1983) *myo*-Inositol, its biosynthesis and metabolism. *Annu. Rev. Plant Physiol.* **34**, 137–161

18. Stephens, L.R. & Irvine, R.F. (1990) Stepwise phosphorylation of *myo*-inositol leading to *myo*-inositol hexakisphosphate in *Dictyostelium*. *Nature (London)* **346**, 580–583

19. Alexandre, J., Lassalles J.P. & Kado R.T. (1990) Opening of Ca^{2+} channels in isolated red beet vacuole membrane by inositol(1,4,5)trisphosphate. *Nature (London)* **343**, 567–570

20. Schroeder, J.I. & Hagiwara, S. (1990) Repetitive increases in cytosolic Ca^{2+} of guard cells by abscisic acid activation of non-selective Ca^{2+} permeable channels. *Proc. Natl. Acad. Sci. U.S.A.* **87**, 9305–9309

4

Artificial cell adhesive proteins

Kiyotoshi Sekiguchi, Toshinaga Maeda and Koiti Titani

Division of Biomedical Polymer Science, Institute for Comprehensive Medical Science, Fujita Health University School of Medicine, Toyoake, Aichi 470-11, Japan

A human body is constructed of approximately 60 billion cells of various kinds. These cells, differing in shape and function, build up an elaborate human body by adhering to each other or to ground substances, called "extracellular matrix", to maintain the homeostasis of life. *Adhesive proteins* mediate such cell-to-cell and cell-to-substrate adhesion under precise control of their temporal and spatial expression.

Adhesive proteins can be classified into two categories: *cell adhesion proteins* mediate cell-to-cell adhesion and *substrate adhesion proteins* mediate cell-to-substrate adhesion[1,2]. The former proteins are membrane proteins, directly bound to cell membranes, which play a major role in tissue-specific intercellular recognition and adhesion. Cadherin and N-CAM (neural-cell adhesion molecule) represent this type of adhesive proteins. On the other hand, most substrate adhesion proteins are major components of the extracellular matrix and participate in regulation of cell growth and differentiation through adhesion to cells. Substrate adhesion proteins are also present in plasma, playing important roles in thrombosis, haemostasis and wound healing. These include fibronectin, laminin, collagen (as extracellular matrix proteins), and vitronectin, von Willebrand factor, and fibrinogen (as plasma proteins).

In this article, we briefly review the structural and functional characteristics of fibronectin, which has been most extensively studied among adhesive proteins, and describe our recent study on engineering of artificial cell adhesive proteins grafted with unique oligopeptide sequences, Arg-Gly-Asp-Ser (RGDS in the one-letter code; see the list of abbreviations at the front of this book) and Glu-Ile-Leu-Asp-Val-Pro-Ser-Thr (EILDVPST), which serve as recognition signals for fibronectin-mediated cell adhesion.

MOLECULAR STRUCTURE AND CELL ADHESIVE SIGNALS OF FIBRONECTIN

Fibronectin is not only one of the major proteins of the extracellular matrix, but it is also present as a soluble protein in plasma at concentrations as high as 300 µg/ml. In addition to its strong cell adhesive activity, it shows a high affinity for binding to other extracellular components such as collagen, fibrin, and heparin.

The molecular structure of fibronectin is shown diagrammatically in Figure 1. Fibronectin is synthesized and secreted as a dimer of similar subunits, each composed of approximately 2500 amino acid residues connected through disulphide bonds located in close proximity to the C-terminal end. After incorporation into the extracellular matrix, the dimeric fibronectin molecules are further cross-linked to each other by additional disulphide bonds.

Each subunit consists of three kinds of *modules* (homologous repeating units) termed types I, II and III[3]. Several of these modules are assembled to form structurally and functionally distinctive domains[4,5]. For example, five type I modules in the N-terminal region of the subunits are assembled to form a domain that is capable of binding heparin, fibrin, and certain types of bacteria, whereas four type I and two type II modules form another domain specifically binding to collagen. Located in the middle of the subunits is the cell-binding domain containing three type III modules, of which the third one contains the active-site RGDS tetrapeptide sequence.

Three of the characteristics identified in the structure of fibronectin, (i) a polymeric structure linked through disulphide bonds, (ii) a modular structure composed of multiple distinct homologous repeating units, and (iii) a multifunctional domain structure, are features also found in among many other substrate adhesion proteins.

In 1984, Pierschbacher & Ruoslahti[6] reported that the four-amino-acid sequence RGDS is the minimal cell adhesive signal of fibronectin. The synthetic RGDS peptide immobilized on plastic or agarose surfaces promotes cell adhesion and spreading. Cell adhesion to fibronectin is completely inhibited by the presence of synthetic oligopeptides containing the RGDS sequence. Replacement of any of the first three residues with other amino acids completely abolishes the activity, but the fourth residue (Ser) can be replaced by Val, Ala, Thr, Cys or Phe. RGDS and equivalent sequences have also been identified in other substrate-adhesion proteins[7]. These proteins include

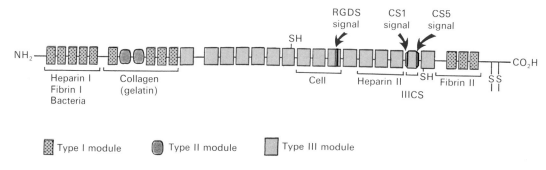

Figure 1. Structure of fibronectin
Fibronectin consists of three types of modules, called type I, type II and type III. These modules are assembled into a series of domains each having specific affinities towards fibrin, heparin, collagen or cell surface receptors, as indicated by brackets.

Figure 2. Structural model of the fibronectin receptor
The fibronectin receptor is composed of two subunits called α and β. The α subunit contains multiple binding sites for calcium ions. The β subunit contains four cysteine-rich repeats (indicated as "Cys"). The cytoplasmic domains of both subunits are linked to the actin cytoskeleton through attachment proteins such as talin, vinculin and α-actinin.

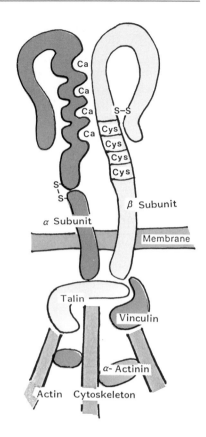

fibrinogen, von Willebrand factor, osteopontin (RGDS sequence), vitronectin (RGDV sequence), laminin (RGDN sequence), thrombospondin and collagen (RGDA sequence).

Besides RGDS, fibronectin contains two other cell adhesive signals with different cell type specificities[8]. These signals are located in the region of the subunits termed IIICS (Figure 1), expression of which is regulated by the alternative pre-mRNA splicing. One of these signals is located in the N-terminal region of IIICS (termed CS1) and displays the specific adhesive activity toward lymphoid cells and certain tumour cells. The activity of the CS1 signal can be reproduced by the synthetic octapeptide EILDVPST. The other signal is located in the C-terminal region of IIICS (termed CS5) and displays adhesive activity toward melanoma cells. The minimal active sequence of the CS5 signal is thought to be Arg-Glu-Asp-Val (REDV).

FIBRONECTIN RECEPTOR ON THE CELL SURFACE

For cells to adhere to a given substrate-adhesion protein, cells should bear a specific receptor for the adhesive protein, e.g. a fibronectin receptor for fibronectin and a laminin receptor for laminin. The adhesiveness of a given cell type to various substrate adhesion proteins depends on the repertoire of cell surface receptors expressed on that particular cell type.

In 1985, Pytela et al.[9] reported the identification and isolation of a fibronectin receptor which consists of two noncovalently associated subunits, called α and β, each having a molecular mass of approximately 140 kDa (Figure 2). The fibronectin receptor is anchored to the actin cytoskeleton, at its cytoplasmic domain, through attachment proteins, such as talin and vinculin. This allows transmission of signals from the extracellular matrix to the cytoplasm, resulting in the reorganization of the actin cytoskeleton necessary for stable cell adhesion. It is probable that, in addition to cell attachment and spreading, cell growth and differentiation are also regulated by the extracellular matrix through these receptors.

Since the discovery and molecular characterization of the fibronectin receptor, many receptor proteins with a high degree of sequence homology to the fibronectin receptor have been identified[10,11]. These include: (i) VLA antigens which appear after 1–2 weeks on activated lymphocytes, (ii) leukocyte adhesion receptors, including the LFA-1, Mac-1, and p150,95 proteins, and (iii) cytoadhesins on platelets and other cell types, such as GPIIb-IIIa and vitronectin receptor (see Table 1). These receptors have a β subunit common to each group (β_1, β_2 and β_3, respectively), but each receptor has a

Table 1. The integrin superfamily

	Other names	α subunits	Ligands
(a) VLA antigens [β_1 (CD29) family]			
VLA1		α_1 (CD49a)	Laminin, collagen
VLA2	GPIa-IIa, ECMR-II	α_2 (CD49b)	Collagen, laminin
VLA3	ECMR-I	α_3 (CD49c)	Fibronectin, laminin, collagen
VLA4	LPAM-2	α_4 (CD49d)	Fibronectin, VCAM-1
VLA5	Fibronectin receptor, ECMR-VI	α_5 (CD49e)	Fibronectin
VLA6	GPIc-IIa	α_6 (CD49f)	Laminin
		α_V (CD51)	Fibronectin, vitronectin?
(b) Leukocyte adhesion receptors [β_2 (CD18) family]			
LFA1		α_L (CD11a)	ICAM-1, ICAM-2
Mac-1	CR3	α_M (CD11b)	C3bi, fibrinogen
p150,95		α_X (CD11c)	C3bi
(c) Cytoadhesins [β_3 (CD61) family]			
Vitronectin receptor		α_V (CD51)	Vitronectin, von Willebrand factor, fibrinogen, thrombospondin
GPIIb-IIIa		α_{IIb} (CD41)	Fibrinogen, fibronectin, vitronectin, von Willebrand factor
(d) Others			
β_4 (CD-?)		α_6 (CD49f)	Epinectin?
β_5 (CD-?)		α_V (CD51)	Vitronectin, fibrinogen
β_P (CD-?)	LPAM-1	α_4 (CD49d)	?

unique α subunit. For example, the fibronectin receptor recognizing the RGDS signal is composed of a β_1 subunit, shared with other VLA antigens, and a unique α_5 subunit. Another receptor for fibronectin, which specifically recognizes the CS1 signal, is also a member of the VLA antigens, but is composed of α_4 and β_1 subunits. Since these receptor proteins integrate the functions of the extracellular matrix and cytoskeleton, they are collectively called *integrins*.

DESIGN OF ARTIFICIAL ADHESIVE PROTEINS

There have been several attempts to make artificial cell adhesive compounds or proteins by utilizing the finding that the RGDS sequence is the minimal structure of the cell adhesive signal of fibronectin. In one such approach, synthetic oligopeptides containing the RGDS sequence have been utilized as specific inhibitors for cell adhesion. Although these oligopeptides possess a low cell adhesive activity, they can inhibit tumour metastasis or platelet aggregation when used at a high dosage. In a second approach, synthetic oligopeptides containing the RGDS sequence are chemically bound to natural or synthetic polymers, to develop new artificial cell adhesive materials. These materials are expected to be useful for treatment of wounds or construction of bioreactors and artificial organs.

As a third approach, we have developed a novel method to produce artificial cell adhesive proteins by grafting the RGDS sequence into non-adhesive proteins by using recombinant DNA techniques[12]. The method has the merits in that (i) such artificial proteins show a much higher cell adhesive potency than synthetic oligopeptides and

Figure 3. Construction of plasmid vectors for expression of Protein A derivatives grafted with the extra tetrapeptides RGDS or RGES
The protein A expression vector pRIT2T contains two restriction sites for *Hind*III (indicated by H in the Figure) within the coding region. The plasmid was partially digested with *Hind*III to linearize at the downstream restriction site which is located in the middle of the coding region. The linearized plasmid was then ligated with double-stranded oligonucleotide cassettes encoding either the RGDS or the RGES tetrapeptide. The oligonucleotide cassettes were designed to create a unique restriction site for *Nhe*I (indicated by N in the Figure) at the 3' end, when inserted in the right orientation. Other unique restriction sites indicated in the figure are: E, *Eco*RI; P, *Pst*I.

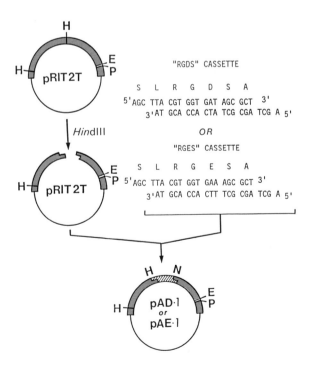

(ii) they may be bifunctional or multifunctional, possessing cell adhesive activity due to the grafted RGDS signal in addition to their original biological activities.

Our newly developed method is briefly described below. We first chose Protein A, the staphylococcal immunoglobulin (IgG)-binding protein, as a model protein to be grafted with RGDS signal for the following reasons: (i) Protein A expressed in bacteria can be easily purified by affinity chromatography on an immobilized IgG column, (ii) an expression vector for Protein A is commercially available, and (iii) the three-dimensional structure of one of the IgG-binding domains of the protein is known, from X-ray crystallographic analysis, so that the conformation surrounding the grafting site is predictable.

We grafted the RGDS sequence into Protein A by *cassette mutagenesis* (Figure 3). A double-stranded oligonucleotide cassette encoding the RGDS sequence (called the RGDS cassette) was chemically synthesized and inserted into the downstream *Hind*III site of the Protein A expression vector, pRIT2T (Pharmacia), using recombinant DNA techniques. As a control, another oligonucleotide cassette encoding a related, but inactive, peptide, RGES, was also synthesized and inserted at the same site on the expression vector. The mutagenized plasmids were introduced to host *Escherichia coli*

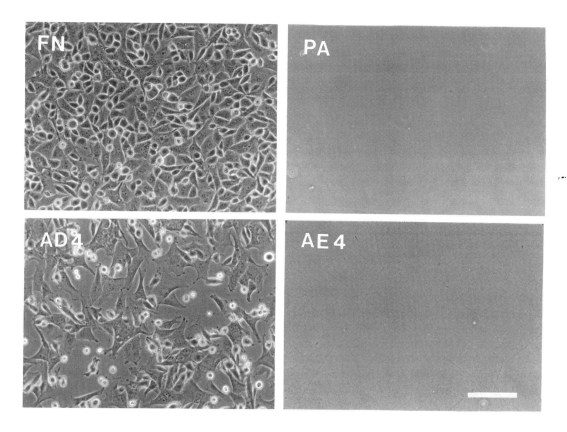

Figure 4. Cell adhesive activity of Protein A derivatives grafted with the RGDS (AD4) or RGES (AE4) sequences
Trypsinized hamster fibroblasts were incubated for 1 hour at 37 °C on polystyrene plates coated with 5 µg of fibronectin/ml (FN), unmodified Protein A (PA), AD4 or AE4. Bar = 100µm.

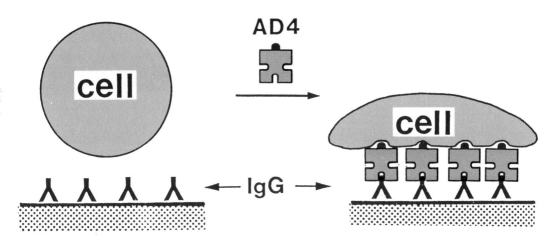

Figure 5. Bifunctionality of RGDS-grafted Protein A
The RGDS-grafted Protein A, AD4, could bridge the cell surface fibronectin receptor and substrate-coated IgG, thereby inducing cell attachment and spreading on the IgG-coated substrate.

and the Protein A derivatives expressed in the host were subsequently purified by affinity chromatography.

CELL ADHESIVE ACTIVITY OF RGDS-GRAFTED PROTEIN A

The cell adhesive activities of the Protein A derivatives thus obtained were examined by a standard cell adhesion assay using hamster kidney fibroblasts. Trypsinized fibroblasts were incubated at 37 °C for 1 hour on polystyrene plates coated with each sample protein, and then the number of cells attached (and spread) to the substrate was counted under a microscope.

As shown in Figure 4, cells attached and spread well on the substrate coated with fibronectin or the mutant Protein A grafted with the RGDS sequence (called AD4) but not on the substrate coated with control Protein A or another mutant Protein A grafted with the RGES sequence (called AE4)[12]. The role of the RGDS sequence in AD4 as an active cell adhesive signal was further demonstrated by the complete inhibition of AD4-mediated cell adhesion by a synthetic RGDS-containing peptide. Furthermore, actin stress fibres were formed in the cells attached and spread on the substrate, indicating that the RGDS signal embedded in Protein A retained an ability to transduce signals to the actin cytoskeleton through its cell surface receptor.

FUNCTIONAL CO-ORDINATION BETWEEN CELL ADHESIVE AND IgG-BINDING ACTIVITIES

Artificial cell adhesive proteins would be worthless if they had lost their original activities as a result of the grafting of the cell adhesive signal. Functional co-ordination of intrinsic IgG-binding activity and the cell adhesive activity, due to the grafted RGDS signal in AD4, was examined by the following assay. Trypsinized hamster fibroblasts were incubated at 37 °C for 1 hour on polystyrene plates first coated with IgG, then treated with each sample protein. As shown in Figure 5, cells attached and spread on the IgG-coated substrate only after pretreatment with AD4, indicating that the bifunctional AD4 can bridge substrate-coated IgG and cell surface RGDS receptors.

PRODUCTION OF PROTEIN A GRAFTED WITH CS1 SIGNAL

As described above, fibronectin contains another cell adhesive signal, CS1, which is specifically recognized by lymphoid cells and tumour cells. In order to assess the transplantability of this signal, we grafted the CS1 signal to Protein A by cassette mutagenesis[13]. The CS1-grafted Protein A showed adhesive activity as high as that of AD4 toward mouse melanoma cells, whereas it showed no activity toward hamster fibroblasts which could attach and spread to AD4. In contrast, lymphoma cells attached to the CS1-grafted Protein A, but not to AD4, indicating that the CS1 signal can also be grafted to Protein A retaining its cell adhesive activity as well as cell type specificity. The CS1-mediated cell adhesion was specifically inhibited by the synthetic EILDVPST octapeptide.

Table 2. Oligopeptide cell adhesive signals

Signals	Origin	Cell types
(a) RGD-related sequences		
RGDS	Fibronectin, von Willebrand factor, osteopontin, fibrinogen	Fibroblasts, many cells
RGDV	Vitronectin	Fibroblasts, endothelial cells
RGDF	Fibrinogen	Platelets
RGDA	Collagen, thrombospondin?	Fibrosarcoma, platelets
RGDN	Laminin	Endothelial cells, neuroblastoma
(b) Heparin-binding sequences		
YEKPGSPPREVVPRPRPGV	Fibronectin	Melanoma cells
RYVVLPRPVCFEKGMNYTVR	Laminin	Melanoma cells, fibrosarcoma, pheochromocytoma
(c) Others		
EILDVPST	Fibronectin	Melanoma cells, lymphocytes
REDV	Fibronectin	Melanoma cells
KQAGDV	Fibrinogen	Platelets
LGTIPG	Laminin	Fibroblasts
IKVAV	Laminin	Pheochromocytoma, melanoma cells, fibrosarcoma
YIGSR	Laminin	Melanoma cells, pheochromocytoma, fibrosarcoma
PDSGR	Laminin	Melanoma cells, fibrosarcoma
HAV	Cadherin	Epithelial cells
YKLNVNDS	gp80	Slime mould

PERSPECTIVES

Many cell adhesive proteins with different specificities have now been identified. As shown in Table 2, they are equipped with RGD or other oligopeptide signals of their own. Besides the RGDS and the CS1 signals of fibronectin, various cell adhesive oligopeptide sequences have been identified in laminin (YIGSR and IKVAV[14]) and fibrinogen (KQAGDV[15]). Furthermore, fibronectin and laminin contain heparin-binding segments enriched in basic amino acid residues, which also serve as adhesive signals. It remains to be tested whether these signals could also be grafted to other proteins, as was the case with the RGDS and CS1 signals of fibronectin.

On the other hand, proteins to be grafted with these signals are not restricted to Protein A. We have already grafted the RGDS signal to calpastatin, a protease inhibitor, to transform it to a cell adhesive protein. Similarly, lysozyme and immunoglobulin can also be conferred with cell adhesive activity by grafting the RGDS sequence. It seems possible, therefore, to endow cell adhesive activity to other proteins, e.g. enzymes, hormones and antibodies, if the grafting site is appropriately selected so that the grafted signal is exposed on the surface of the recipient protein with a minimal distortion of the original conformation.

Such signal grafting is unlimited in the choice of adhesive signals and recipient proteins, allowing production of a wide variety of artificial, multifunctional adhesive proteins with defined specificities and additional biological activities associated with recipient proteins. Artificial cell adhesive proteins thus created may well be useful not only as materials for development of bioreactors or artificial organs, but also for targeting of protein drugs, accelerating wound healing, and preventing thrombosis and tumour metastasis.

REFERENCES

1. Edelman, G. (1986) Cell adhesion molecules in the regulation of animal form and tissue pattern. *Annu. Rev. Cell Biol.* **2**, 81–116

2. Takeichi, M. (1988) The cadherins: cell-cell adhesion molecules controlling animal morphogenesis. *Development* **102**, 639–655

3. Petersen, T.E., Skorstengaard, K. & Vibe-Pedersen, K. (1989) in *Fibronectin* (Mosher, D.F., ed.), pp.1–24, Academic Press, San Diego

4. Sekiguchi, K. & Hakomori, S. (1983) Domain structure of human plasma fibronectin. *J. Biol. Chem.* **258**, 3967–3973

5. Yamada, K.M. (1989) in *Fibronectin* (Mosher, D.F., ed.), pp.47–121, Academic Press, San Diego

6. Pierschbacher, M.D. & Ruoslahti, E. (1984) Cell attachment activity of fibronectin can be duplicated by small synthetic fragments of the molecule. *Nature (London)* **309**, 30–33

7. Ruoslahti, E. (1988) Fibronectin and its receptors. *Annu. Rev. Biochem.* **57**, 375–413

8. Humphries, M.J. & Yamada, K.M. (1990) in *Morphoregulatory Molecules* (Edelman, G.M., Cunningham, B.A. & Thiery, J.P., eds.), chapter 6, pp.137–172, John Wiley, Chichester

9. Pytela, R., Pierschbacher, M.D. & Ruoslahti, E. (1985) Identification and isolation of a 140 kd cell surface glycoprotein with properties expected of a fibronectin receptor. *Cell* **40**, 191–198

10. Albelda, S.M. & Buck, C.A. (1990) Integrins and other cell adhesion molecules. *FASEB J.* **4**, 2868–2880

11. Springer, T.A. (1990) The sensation and regulation of interactions with the extracellular environment: the cell biology of lymphocyte adhesion receptors. *Annu. Rev. Cell Biol.* **6**, 359–402

12. Maeda, T., Oyama, R., Ichihara-Tanaka, K., Kimizuka, F., Kato, I., Titani, K. & Sekiguchi, K. (1989) A novel cell adhesive protein engineered by insertion of the Arg-Gly-Asp-Ser tetrapeptide. *J. Biol. Chem.* **264**, 15165–15168

13. Maeda, T., Titani, K. & Sekiguchi, K., unpublished work

14. Tashiro, K., Sephel, G.C., Weeks, B., Sasaki, M., Martin, G.R., Kleinman, H.K. & Yamada, Y. (1989) A synthetic peptide containing the IKVAV sequence from the A chain of laminin mediates cell attachment, migration, and neurite outgrowth. *J. Biol. Chem.* **264**, 16174–16172

15. Tranqui, L., Andrieux, A., Hudry-Clergion, G., Ryckewaert, J.-J., Soyez, S., Chapel, A., Ginsberg, M.H., Plow, E.F. & Marguerie, G. (1989) Differential structural requirements for fibrinogen binding to platelets and to endothelial cells. *J. Cell Biol.* **108**, 2519–2527

5

The urea cycle: a two-compartment system

Malcolm Watford

Department of Nutritional Sciences, Cook College, Thompson Hall, Rutgers, The State University of New Jersey, New Brunswick, NJ 08903, U.S.A.

INTRODUCTION

Urea was first isolated from urine by Roulle in 1773 as a "substance savonneuse" which yielded carbonic acid and ammonia. Davy probably synthesized urea in 1812, but he failed to recognize it as such so that the credit for the synthesis of urea has gone to Wohler. In 1828, in an experiment known to all students of chemistry, Wohler synthesized urea (an organic compound) from ammonia and lead cyanate (inorganic precursors). During the next 100 years many theories for the biosynthesis of urea were proposed, often involving ammonium cyanate as an intermediate. However, these theories were ruled out 60 years ago when Hans Krebs and Kurt Henseleit carried out the experiments providing evidence for the first description of a metabolic cycle, the urea cycle[1]. Using liver slices in incubation, Krebs and Henseleit found that ornithine stimulated urea synthesis without itself being utilized in the process. From this they suggested a cyclic process involving ammonia combining with ornithine to give citrulline. This citrulline then gives rise to arginine which is then hydrolysed to urea with the regeneration of the ornithine. The original proposed cycle was

---

Dedicated to the memory of Professor Sir Hans Krebs, F.R.S., to commemorate his experiments 60 years ago resulting in the formulation of the urea cycle.

Ammonia refers to the sum of NH_3 plus NH_4^+. Where a specific molecular species is important it is given in parentheses.

incomplete and others have filled in the details (Figure 1), but the basic concept has stood the test of time.

Although strictly speaking only four enzymes are involved in the cycle proper, the necessity for carbamoyl phosphate means that carbamoyl phosphate synthetase 1 is usually described as a urea cycle enzyme. As shown in Figure 2, the pathway begins in the mitochondria with ammonia (NH_3) and bicarbonate combining via carbamoyl phosphate synthetase 1 to give carbamoyl phosphate, which then combines with ornithine to give citrulline via the ornithine transcarbamoylase reaction. The citrulline then exits from the mitochondria to the cytosol where nitrogen is added from aspartate through the action of argininosuccinate synthase. Fumarate is then removed by argininosuccinate lyase and the arginine so formed is hydrolysed by arginase to give urea and ornithine.

Although the pathway is universally accepted, unanswered questions remain concerning its regulation and even its physiological function. The cycle is usually described as a detoxification mechanism for ammonia, but it also removes bicarbonate and is therefore directly linked to acid-base homeostasis. Since amino acid catabolism generates both ammonia and bicarbonate the debate has centred on which function, removal of ammonia or bicarbonate, is of primary importance. Careful evaluation of the evidence shows however, that under conditions where ammonia levels rise, ammonia detoxification is the driving force for urea synthesis[2,3]. Furthermore, changes in the rate of urea synthesis appear to be of little consequence to acid-base balance *in vivo*[3,4]. The reader is referred to the reviews of Atkinson for the bicarbonate argument and those of Walser and Brosnan *et al.* for the ammonia argument. A more balanced and succinct commentary is to be found in the review by Meijer *et al.*[2].

DISTRIBUTION

The cycle is found in all ureotelic animals but it is also present in some non-ureotelic species, predominantly the elasmobranchs (cartilaginous fish such as sharks, rays, skates), where it functions to provide urea for use as an osmotic regulator. In mammals the complete cycle is found only in the liver but partial reactions are found in certain other tissues. Carbamoyl phosphate synthetase 1 and ornithine transcarbamoylase are present in the intestinal mucosa, while argininosucinate synthase and argininosuccinate lyase are found in the kidney. This constitutes a tissue-compartmented pathway with ornithine being converted to citrulline in the intestine and then modified in the kidney to yield arginine[5]. This arginine is then available for the protein synthesis and anaplerotic functions throughout the body.

REGULATION

In mammalian liver the cycle is subject to both short-term (changes in the specific activity of existing enzyme protein) and long-term (changes in the amount of enzyme protein) regulation. The first real evidence of regulation came from the experiments of Folin[6] who, by feeding himself a low-protein diet of arrowroot starch and cream, showed that urea production was related to dietary protein load. This was followed in the 1930s by the demonstration that arginase activity varied with dietary protein level and in the 1960s Schimke showed co-ordinate regulation of the five enzymes in response to dietary protein intake and during starvation. In fact, he used these

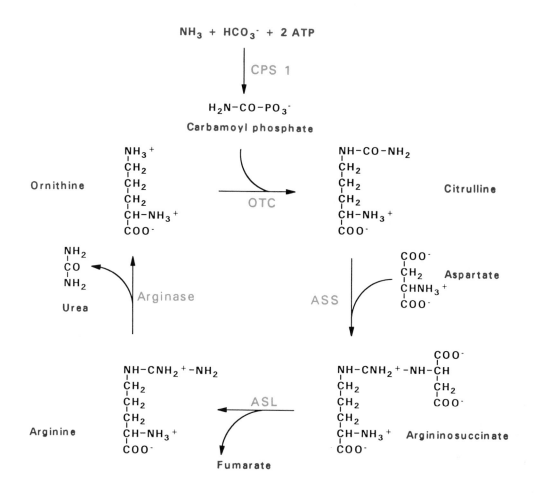

Figure 1. The urea cycle
Full details of the enzymes and compartmentation are given in the text. Abbreviations: CPS1, carbamoyl phosphate synthetase 1; OTC, ornithine transcarbamoylase; ASS, argininosuccinate synthetase; ASL, argininosuccinate lyase.

responses as models to delineate the quantitative aspects of protein synthesis and degradation[7]. Although changes in the activity of all five enzymes appear to be co-ordinately regulated, the molecular mechanisms are not always the same for each enzyme. Regulation occurs at one or more of a number of sites, from protein synthesis and degradation to changes in gene transcription[8].

The urea cycle enzymes have quite long apparent half-lives (of the order of days) and therefore the long-term changes in enzyme concentration take some time to be manifested. This may mean that such changes are simply adaptive in nature and do not play a regulatory role. This may be especially true in subjects eating a regular diet. However, there is a need for short-term regulation which is evident from the knowledge that the urea cycle does not utilize all of the substrate available to it.

Since ammonia is an important intermediate in other pathways, such as the synthesis of pyrimidines, glutamate and other amino acids, the concentration of ammonia in the liver is held relatively constant. Similarly, the concentration of bicarbonate within the liver cell does not vary significantly.

The urea cycle enzymes all work *in vivo* at substrate concentrations below their K_m values. Therefore teleologically it would make sense to prevent ammonia from entering the cycle, which suggests carbamoyl phosphate synthetase 1 as the site of regulation. According to the control strength theory of regulation the flux control coefficient of carbamoyl phosphate synthetase 1 is close to 1, indicating it is the major regulatory step and it is the only urea cycle enzyme known to be subject to allosteric regulation[2]. Indeed, it has an absolute requirement for the regulator N-acetylglutamate, and although controversial there is now sound evidence that the levels of N-acetylgluta-mate parallel the rate of flux through the cycle and play an important regulatory role. A recent review gives an excellent overview of the short-term regulation of the urea cycle by N-acetylglutamate and substrate availability[2].

COMPARTMENTATION

As indicated above and outlined in Figure 2, the urea cycle involves reactions in two classical compartments within the liver cell, the mitochondrion and the cytosol. But the cycle is subject to further compartmentation. First, there is hepatic hetero-geneity: not all hepatocytes have equal activities of the urea cycle enzymes. Higher levels of the enzymes for the urea cycle, gluconeogenesis and amino acid catabolism are found in periportal cells while glutamine synthetase activity is located exclusively in perivenous cells. This has given rise to an intrahepatic–intercellular system for ammonia detoxification, by either urea synthesis in periportal cells, or glutamine synthesis in perivenous cells[2].

MICRO-COMPARTMENTATION AND CHANNELLING

In formulating models of metabolic regulation the court of last resort must always be the intact animal, but it is impossible to obtain sufficient information from this source. Instead we look at experiments using a variety of techniques *in vitro*, prep-arations of varying degrees of purity, isolated organs, cells, organelles and enzymes. Given the problems and compromises inherent in such systems it is perhaps surprising that the results obtained have usually been consistent with observations *in vivo*. For example, despite the knowledge of differences in the amount of urea cycle enzymes in different hepatocytes, the gradient of enzyme activity across the liver acinus is not usually considered of functional importance in studies of isolated hepatocytes.

Such models of regulation that have been proposed often acknowledge that the properties of enzymes *in vitro* may only partially resemble those *in vivo*, and sometimes there is direct mention of problems with compartmentation. Usually this refers to difficulties in obtaining values for the true concentrations of intermediates at the local environment of the enzyme due to membrane barriers, or metabolite binding. Despite these somewhat obvious limitations the idea is still fostered in textbooks that enzymes in solution behave essentially as they do in the intact cell and that both enzymes and intermediates are in some way "soluble". In recent years this simple notion has been questioned; enzyme (protein) concentration in the cell is clearly much higher than in enzyme assays *in vitro*. Furthermore, there are questions of solvation capacity, enzyme–enzyme interactions, and localization of enzymes within the cell. Experiments with tracers have shown non-mixing of pathway intermediates with other metabolite pools, and mutants lacking just one enzyme of a pathway often fail to carry out any of the other reactions. Such findings have led to suggestions that many pathways are organized in some form of loose enzyme–enzyme associations and that intermediates are passed along the chain of enzymes, a process known as substrate (metabolite) channelling. Srere has reviewed this subject in detail and has coined the word "metabolon" for such complexes[9].

The urea cycle is a pathway where all five of the urea cycle enzymes have been described as soluble, even those located within the mitochondrial matrix, but there is now extensive evidence that they are not freely diffusible and that considerable channelling of intermediates may occur.

LOCATION OF ENZYMES

The mitochondrial enzymes, carbamoyl phosphate synthetase 1 and ornithine transcarbamoylase, are sometimes described as "in loose association with the inner mitochrondrial membrane". This is because treatment with mild detergent is required to completely "solubilize" the activities. Recently, immunoelectron microscopy has demonstrated that 41% of carbamoyl phosphate synthetase 1 and 59% of ornithine

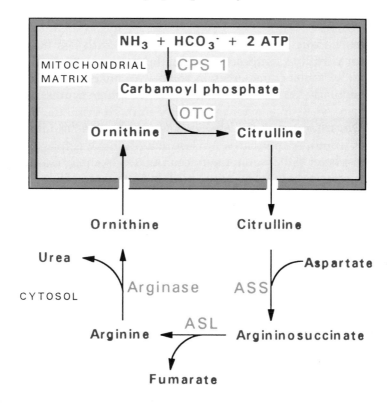

Figure 2. Compartmentation of the urea cycle
Abbreviations for enzyme names are explained in the caption to Figure 1.

transcarbamoylase are located within 15 nm of the cristae, indicating very close association with the membrane[10,11]. This localization would favor direct transfer of carbamoyl phosphate to ornithine transcarbamoylase and production of citrulline.

Similarly, a certain proportion of liver arginase is always found associated with isolated hepatic mitochondria. This is not simply an artifact of isolation (often proteins will stick to membranes during the homogenization procedure) since the arginase is not removed by quite high (150mM-KCl) salt washes[12]. Although this arginase only represents about 10% of the total hepatic arginase activity, this is more than that required for maximal urea cycle capacity.

CHANNELLING WITHIN THE MITOCHONDRION

Experiments with labelled substrates and intermediates have indicated that various parts of the cycle are organized in such a way as to result in channelling. When mitochondria isolated from rat liver are incubated with radiolabelled ornithine this ornithine is utilized for citrulline synthesis in preference to ornithine generated within the mitochondrial matrix. In such experiments ornithine can only be detected in the mitochondrial matrix after incubation with very high concentrations of extramitochondrial ornithine. When these ornithine-loaded mitochondria are incubated with radiolabelled ornithine the citrulline formed has the same specific radioactivity as the extramitochondrial ornithine. Such results indicate preferential use of the extramitochondrial radiolabelled ornithine and little mixing with the intra-matrix unlabelled pool[13].

When isolated hepatic mitochondria are incubated with less than maximal levels of substrates for citrulline synthesis then ornithine transport into the mitochondrion slows. This indicates that transport is in some way linked to further metabolism of the ornithine. Similarly, flux through carbamoyl phosphate synthetase 1 requires ornithine transcarbamoylase in a manner that flux through ornithine transcarbamoylase is essential, since carbamoyl phosphate never accumulates in the mitochondrial matrix. A deficiency of ornithine transcarbamoylase activity does not result in a buildup of carbamoyl phosphate; rather, ammonia accumulates (some carbamoyl phosphate is degraded and some is diverted to orotic acid, but these are of very small magnitude when compared to flux through the urea cycle).

The question of the type of assocation between enzymes can also be addressed by looking at the carbamoyl phosphate synthetase 1–ornithine transcarbamoylase system. Carbamoyl phosphate synthetase 1 is one of the most abundant hepatic mitochondrial proteins; it comprises up to 26% of matrix protein according to some accounts. However, the level of ornithine transcarbamoylase protein in the matrix is of the order of 1/200 that of carbamoyl phosphate synthetase 1, yet because of differences in specific activity the total activity of ornithine transcarbamoylase is 200-fold higher than that of carbamoyl phosphate synthetase 1. These numbers themselves mean that all carbamoyl phosphate in the matrix is bound to an enzyme. Furthermore, work with the sparse fur ash mouse (ornithine transcarbamoylase deficiency with abnormal skin and hair) has shown that hepatic mitochondria with only 5% of the normal ornithine transcarbamoylase activity can efficiently utilize all the carbamoyl phosphate provided[14]. Such data support channelling of carbamoyl phosphate but exclude a direct stochiometric association of the two enzymes.

CHANNELLING IN THE CYTOSOL

Evidence of channelling within the mitochondrial matrix and linkage to ornithine transport is essentially a one-compartment system. Recently Raijman and colleagues have extended their work to demonstrate channelling in the remaining, cytosolic, part of the cycle[15]. They used rat hepatocytes treated with alpha-toxin, which forms small holes in the membrane, incubated for short periods in a medium resembling the intracellular milieu. Small molecules (amino acids) can penetrate into such cells but larger molecules (e.g. inulin, 5 kDa) are excluded. For example, the intermediates of the urea cycle, citrulline and argininosuccinate, cannot penetrate into intact cells but in the permeabilized cells they enter and share the same space as water.

These cells were incubated with $H^{14}CO_3^-$ which results in the synthesis of labelled urea, the carbon being incorporated at the carbamoyl phosphate synthetase 1 step. Raijman and colleagues reasoned that if unlabelled citrulline, argininosuccinate or arginine were added to such cells, and if the exogenous intermediates were free to mix with the endogenous pool, then the amount of radioactivity incorporated into urea should be decreased. In practice, the presence of 200-fold excess arginine (2mM) did not decrease the percentage or the total amount of ^{14}C found in urea. Similarly, there was no increase in the amount of ^{14}C found in arginine, despite the fact that large amounts of unlabelled arginine were hydrolysed within the cell. In order to decrease the amount of ^{14}C in urea it was necessary to use extremely high levels of arginine (5mM). Even then only a 50% decrease in radioactivity was obtained and this was probably due to arginine inhibition of argininosuccinate synthase, as evidenced by the recovery of large amounts of ^{14}C in citrulline, rather than dilution of the arginine pool. The results indicate that exogenous arginine enters the permeabilized cells where it is available to considerable arginase activity but it is not freely mixing with arginine derived endogenously from mitochondrial citrulline.

When such experiments were tried with unlabelled citrulline or argininosuccinate similar results were obtained. Less tight channelling (more dilution of the radioactivity) was found with citrulline but the amount of ^{14}C in urea was never reduced by more than 50% with 2mM unlabelled intermediates. These results indicate that free mixing of the exogenous intermediates with the endogenously generated labelled intermediates had not occurred; if it had, the label in urea would have been diluted to below the level of detection. The conclusion drawn from these studies is that citrulline released from the mitochondria is efficiently passed to argininosuccinate synthase, the argininosuccinate to argininosuccinate lyase and finally the arginine to arginase[15].

CHANNELLING IN THE COMPLETE CYCLE?

The evidence from enzyme localization, metabolite transport, pathway flux and labelling experiments all indicate that the urea cycle operates as a metabolon. There seems to be efficient channelling between ornithine in the cytosol, carbamoyl phosphate synthetase 1, ornithine transcarbamoylase, possibly citrulline transport out of the mitochondria (citrulline never accumulates in the matrix) and the cytosolic portion of the pathway from citrulline to urea. Given the location of considerable arginase activity in association with the mitochondria it is also possible that this results in channelling of ornithine from arginase directly into the mitochondrion, thus completing the cycle.

The concept of a metabolic pathway is artificial; it is certainly convenient to view the urea cycle as beginning at carbamoyl phosphate synthetase 1 and ending with urea. In reality, however, no pathway exists in isolation from other processes and we must consider the source of substrates and the fate of the products. At least the urea cycle allows easy identification of the latter: the urea is excreted. But where do the ammonia, bicarbonate and aspartate come from, and are they channelled into the cycle?

Work with intact hepatocytes has demonstrated that aspartate produced within the mitochondrion is preferentially used by the cytosolic enzyme argininosuccinate synthase[16]. Although not proof of channelling, this does show selective use of intermediates.

If the enzymes of amino acid degradation are considered as part of the urea cycle pathway then extensive evidence of enzyme–enzyme associations and even channelling have been reported. Glutamate dehydrogenase, carbamoyl phosphate synthetase 1, ornithine transcarbamoylase and mitochondrial aspartate aminotransferase co-precipitate in polyethylene glycol, and this is interpreted as evidence of association *in vivo* which increases the potential for channelling[17]. Meijer[18] found that the ammonia released from glutamine via glutaminase is channelled into the carbamoyl phosphate synthetase 1 reaction and thereby to urea.

SUMMARY

Channelling and enzyme localization remain controversial; they have been proposed for a number of metabolic pathways, especially glycolysis and the tricarboxylic acid cycle. One aspect that is often overlooked in discussions of channelling is that very tight channelling is readily accepted in pathways which occur in enzyme complexes, for example fatty acid synthetase, the 2-oxoacid dehydrogenase complexes and the protein synthesis/ribosome complex.

As a metabolon the urea cycle is presently unique since it covers two conventional compartments. For the urea cycle, channelling appears to be almost complete with varying degrees of tightness at each step. Since a considerable portion of the nitrogen for urea synthesis is derived within the hepatocyte from amino acids, does this means that numerous enzyme–enzyme associations are required for this metabolon? Another important, as yet unaddressed, question is what are the consequences of channelling to theories of metabolic regulation? The answers will no doubt be forthcoming in the next few years as the concept of metabolons gains or loses acceptance.

REFERENCES

1. Krebs, H.A. & Henseleit, K. (1932) Untersuchungen uber die harnstoffbildung im tierkorper. *Hoppe-Seyler's Z. Physiol. Chem.* **210**, 33–66

2. Meijer, A.J., Lamers, W.H. & Chamuleau, R.A.F.M. (1990) Nitrogen metabolism and ornithine cycle function. *Physiol. Rev.* **70**, 701–749

3. Almdal, T., Vitstrup, H., Bjerrum, K. & Kristensen, L.O. (1989) Decrease in ureagenesis by partial hepatectomy does not influence acid base balance. *Am. J. Physiol.* **258**, F691–F696

4. Bjerrum, K., Vilstrup, H., Almdal, T.P. & Kristensen, L.O. (1990) No effect of bicarbonate induced alkalosis on urea synthesis in normal man. *Scand. J. Clin. Lab. Invest.* **50**, 137–141

5. Dhanakoti, S.N., Brosnan, J.T., Herzberg, G.R. & Brosnan, M.E. (1990) Renal arginine synthesis: studies *in vitro* and *in vivo*. *Am. J. Physiol.* **259**, E437–E442

6. Folin, O. (1905) Laws governing the chemical composition of urine. *Am. J. Physiol.* **13**, 66–115

7. Schimke, R. & Doyle, D. (1970) Control of enzyme levels in animal tissues. *Annu. Rev. Biochem.* **39**, 929–976

8. Morris, S.M., Moncman, C.L., Rand, K.D., Dizikes, G.J., Cederbaum, S.D. & O'Brien, W.E. (1987) Regulation of mRNA levels for five urea cycle enzymes in rat liver by diet, cyclic AMP, and glucocorticoids. *Arch. Biochem. Biophys.* **256**, 343–353

9. Srere, P.A. (1987) Complexes of sequential metabolic enzymes. *Annu. Rev. Biochem.* **56**, 89–124

10. Powers-Lee, S.G., Mastico, R.A. & Bendayan, M. (1987) The interaction of rat liver carbamoyl phosphate synthetase and ornithine transcarbamoylase with inner mitochondrial membranes. *J. Biol. Chem.* **262**, 15683–15688

11. Yokota, S. & Mori, M. (1986) Immunoelectron microscopical localization of ornithine transcarbamylase in hepatic parenchymal cells of the rat. *Histochem. J.* **18**, 451–457

12. Cheung, C.-W. & Raijman, L. (1981) Arginine, mitochondrial arginase, and the control of carbamyl phosphate synthesis. *Arch. Biochem. Biophys.* **209**, 643–649

13. Cohen, N.S., Cheung, C.-W. & Raijman, L. (1987) Channeling of extramitochondrial ornithine to matrix ornithine transcarbamoylase. *J. Biol. Chem.* **262**, 203–208

14. Cohen, N.S., Cheung, C.-W. & Raijman, L. (1989) Altered enzyme activities and citrulline synthesis in liver mitochondria from ornithine carbamoyltransferase-deficient sparse-fur ash mice. *Biochem. J.* **257**, 251–257

15. Cheung, C.-W., Cohen, N.S. & Raijman, L. (1989) Channeling of urea cycle intermediates *in situ* in permeabilized hepatocytes. *J. Biol. Chem.* **264**, 4038–4044

16. Meijer, A.J., Gimpel, J.A., Deleeuw, G., Tischler, M.E., Tager, J.M. & Williamson, J.R. (1978) Interrelationships between gluconeogenesis and ureogenesis in isolated hepatocytes. *J. Biol. Chem.* **253**, 2308–2320

17. Fahien, L.A., Kmiotek, E.H., Woldegiorgis, G., Evenson, M., Shrago, E. & Marshall, M. (1985) Regulation of aminotransferase-glutamate dehydrogenase interactions by carbamoylphosphate synthase-1, Mg^{2+} plus leucine versus citrate and malate. *J. Biol. Chem.* **260,** 6069–6079

18. Meijer, A.J. (1985) Channeling of ammonia from glutaminase to carbamoyl-phosphate synthetase in liver mitochondria. *FEBS Lett.* **191**, 249–251

SUGGESTIONS FOR FURTHER READING

History of the urea cycle

Krebs, H.A. (1982) The discovery of the ornithine cycle of urea synthesis. *Trends Biochem. Sci.* **7**, 76–78

Holmes, F.L. (1980) Hans Krebs and the discovery of the ornithine cycle. *Fed. Proc.* **39**, 216–225

Grisolia, S., Baguena, R. & Mayor, F. (eds.) (1976) The Urea Cycle, John Wiley & Sons, New York

Evolution and function of the urea cycle

Nener, J.C. (1988) Variable and constrained features of the ornithine-urea cycle. *Can. J. Zool.* **66**, 1069–1077

Takiguchi, M., Matsubasa, T., Amaya, Y. & Mori, M. (1989) Evolutionary aspects of urea cycle enzyme genes. *BioEssays* **10**, 163–166

Atkinson, D.E. & Bourke, E. (1984) The role of ureagenesis in pH homeostasis. *Trends Biochem. Sci.* **9**, 291–302

Brosnan, J.T., Lowry, M., Vinay, P., Gougoux, A. & Halperin, M.L. (1987) Renal ammonium production — une vue canadienne. *Can. J. Physiol. Pharmacol.* **65**, 489–498

Walser, M. (1986) Roles of urea production, ammonium excretion, and amino acid oxidation in acid–base balance. *Am. J. Physiol.* **250**, F181–F188

Short-term regulation

Meijer, A.J., Lamers, W.H. & Chamuleau, R.A.F.M. (1990) Nitrogen metabolism and ornithine cycle function. *Physiol. Rev.* **70**, 701–749

Long-term regulation

Morris, S.M., Moncman, C.L., Rand, K.D., Dizikes, G.J., Cederbaum, S.D. & O'Brien, W.E. (1987) Regulation of mRNA levels for five urea cycle enzymes in rat liver by diet, cyclic AMP, and glucocorticoids. *Arch. Biochem. Biophys.* **256**, 343–353

Jackson, M., Beaudet, A. & O'Brien, W.E. (1986) Mammalian urea cycle enzymes. *Annu. Rev. Genet.* **20**, 431–464

Channelling

Srere, P.A. (1987) Complexes of sequential metabolic enzymes. *Annu. Rev. Biochem.* **56**, 89–124

Spivey, H.O. & Merz, J.M. (1989) Metabolic compartmentation. *BioEssays* **10**, 127–130

Maretzki, D., Reimann, B. & Rapoport, S.M. (1989) A reappraisal of the binding of cytosolic enzymes to erythrocyte membranes. *Trends Biochem. Sci.* **14**, 93–96

6

Antibody engineering: an overview

Richard O'Kennedy and Paul Roben

School of Biological Sciences, Dublin City University, Dublin 9, Ireland

INTRODUCTION

Antibodies are amongst the most exquisitely engineered molecules in nature. They have a common basic structure and function, but their genetics and production are quite complex and our understanding of the mechanisms involved is still far from complete. However, in this Essay we will show how the basic structure and functions of antibodies are related and how antibody engineering (genetic, chemical and cell) may help to elucidate their mode of action and expand and improve their applications both *in vitro* and in *vivo*.

The basic structure of an antibody molecule is shown in Figure 1. An antibody consists of two heavy (H) and two light (L) chains [either κ (kappa) or λ (lambda)] which are joined together by disulphide linkages[1]. There is a hinge region in the heavy chains which facilitates conformational changes in the antibody which occur during circulation in the organism or following binding to antigens. The binding site of the antibody is situated in the variable region at the -NH$_2$ end of both the H and L chains. The constant regions of the antibody chain provide a backbone structure. The specificity for antigen binding is determined by the three-dimensional structure of the antibody combining site and the precise interaction depends on the amino acid residues which react with the antigen. In the heavy chain the CH$_1$, CH$_2$ and CH$_3$ constant domains have specific functions. For example, CH$_1$ is involved in binding the C4b fragment of complement, the CH$_2$ region has carbohydrate moieties attached (which may play a role in secretion, protection against proteolytic attack and maintaining antibodies in circulation in the blood) and it fixes the C1q unit of complement, and CH$_3$ binds to Fc receptors on cells such as macrophages and monocytes[1]. Anti-

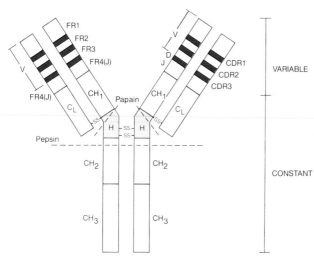

Figure 1. Diagram of antibody domains
The variable and heavy chains of a complete antibody are shown. The positions of the complementarity determining regions (CDRs, red) and hinge region (pink) are also indicated. The broken lines mark the positions at which pepsin and papain cut antibodies.

Table 1. Properties of major immunoglobulin classes

	IgG	IgA	IgM	IgD	IgE
Molecular mass	150 kDa	160 kDa (320 kDa dimer)	900 kDa (pentamer)	185 kDa	200 kDa
Number of units	1	1 (2)	5	1	1
Heavy chains	ω	α	μ	θ	ε
Antigen binding sites available	2	2 (4)	5 (10)	2	2
Approximate proportion of total antibody	80%	13%	6%	0.1%	0.002%
Major biological properties and roles	Combats micro-organisms Crosses placenta Fixes complement Binds to phagocytic cells Major antibody of secondary response	Major antibody in sero-mucous secretions Defends external body surfaces	Major agglutinating antibody Antibacterial Major primary antibody in immune response	Present on lymphocyte surface	Levels raised in parasitic infections Major role in many allergic responses

bodies can be divided into classes and subclasses, each with their own forms and specific functions (as shown in Table 1).

The light and heavy chain variable regions (V_L and V_H, respectively) consist of a highly conserved framework structure linked by three hypervariable loops (also known as complementarity-determining regions, CDRs) on each of the V_L and V_H chains. The framework is made up of two β-pleated sheets which are linked by the three loops making up the hypervariable regions[2] (Figure 2). The total length of each chain is approximately 110 amino acids. The three hypervariable loops are located approximately at positions 26 to 33, 50 to 53 and 91 to 96.

Antibody specificity is defined, in part, by the amino acid composition of the hyper-variable loop regions. However, it is now thought that there are a limited number of conformations available to the loops in any antibody[2,3]. These conformations are known as canonical structures and are determined by amino acids at certain key points within the loops. These amino acids may only play a role in loop structure and not in antibody specificity. The remaining residues within the loops determine specificity. It is apparent, therefore, that although antibodies may vary greatly in specificity, the variable regions are very highly defined structurally.

Antibody-derived fragments may be produced by a number of chemical methods and proteolytic cleavage. Two enzymes commonly used to produce antibody fragments are pepsin and papain (see Figure 1). Pepsin produces an F(ab')$_2$ fragment and an Fc fragment, while papain produces two Fab fragments and an Fc fragment. Originally their production was designed to help elucidate antibody structure. However, they were found to have inherent advantages in some applications. In Figure 3 the portions of the antibody molecule which make up the Fab and F(ab')$_2$ units are shown. Fv fragments can also be produced, which consist of a variable heavy chain (V_H) and a variable light chain (V_L) in association, to give a complete binding site. The advantages of using these fragments is that they may overcome problems associated with the presence of the constant heavy region (Fc), such as non-specific binding and reduced access to tissue sites due to the larger size of the whole antibody unit.

It is also possible to generate bifunctional antibodies having different specificities on either arm of the antibody, as shown in Figure 4. For example, such antibodies

Figure 2. Arrangement of the CDR loops and framework regions in V_L and V_H

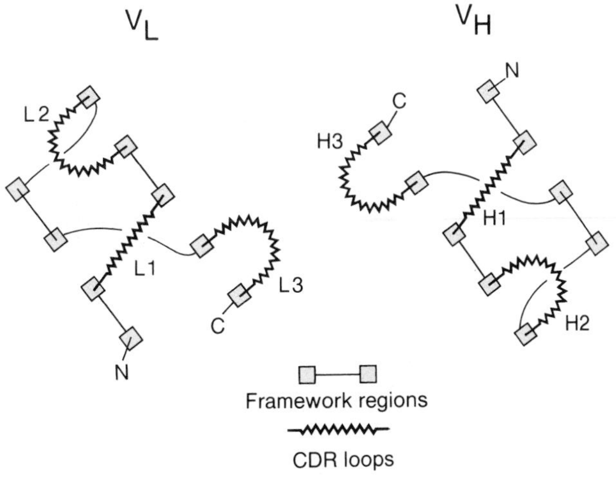

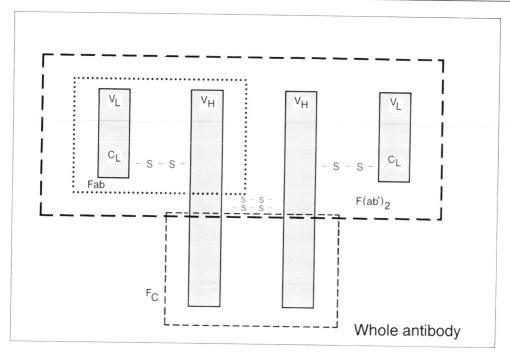

Figure 3. The structure of an antibody and its Fab, F(ab')₂ and Fc fragments

could be used to target therapeutic substances, or cells of the immune system, to tumours. The advantage here is that for such binding neither the antibody or the antigen need be chemically modified, an approach which can alter both antibody and drug efficiency[4].

Antibodies will be produced if a host is exposed to a foreign molecule or antigen. The presence of the antigen stimulates the development and expansion of a clone (or clones) of cells secreting specific antibody to the antigen. The process involved in very complex, involving T cells, B cells and macrophages and a host of cytokines and other factors[1]. A major advance in the production of antibodies was the development of monoclonal antibodies by Kohler and Milstein in 1975. This allowed the

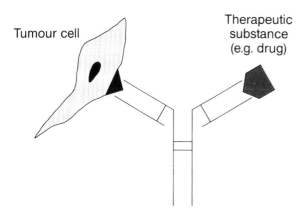

Figure 4. A bifunctional antibody with one binding site reactive with a tumour cell surface antigen and the other site binding to a therapeutic substance

fusion of antigen-sensitized spleen cells with myeloma cells to produce hybridomas secreting antibodies of the required specificity. However, while this approach allowed the generation of murine antibodies with the required characteristics and greatly expanded the applications of antibody technology for diagnosis and therapy, it has certain attendant limitations. This has led to the development of other techniques for producing antibodies, such as genetic engineering.

WHY ANTIBODY ENGINEERING?

The advent of genetic engineering has provided a powerful tool to study antibodies and their properties[5–8]. It has the advantage of being able to alter genes precisely to produce antibodies with required specificities and improved affinities. The following summarizes the applications of antibody engineering:

1. Production of humanized antibodies from existent murine antibodies.

2. Production of active antibody derivatives and fragments with optimized properties for particular applications, e.g. bifunctional antibodies, Fab, F(ab')$_2$, Fv or single-domain antibodies.

3. Studies on the binding sites of antibodies and the process of molecular recognition.

4. Generation of antibodies with higher specificity and affinity.

5. Production of a range of hybrid proteins having antigen binding properties and additional properties such as enzymic activity, or toxic effects due to the presence of a genetically linked toxin chain.

6. Development of a greater efficiency in the large scale production of antibodies.

In order to understand the mechanisms involved in the production of antibodies in the cell, a review of the genetics of antibody production is necessary.

THE GENETICS OF ANTIBODY DIVERSITY

The human body can produce over 200 million antibodies, enough for every known antigen it may encounter[1]. What is the genetic basis of such a vast diversity of antibody molecules? At first, two theories were postulated to explain this. The germ line theory suggested that there was a different gene for every antibody light and heavy chain, whereas the somatic theory states that there are only a small number of heavy and light chain genes which would be rearranged after stimulation by the revelant antigen. Since the germ line theory would require too large a section of the genome to be devoted to antibody expression, the latter theory was taken to be correct, although certain aspects of the germ line theory also aid in understanding antibody diversity.

It has been shown that light chains can be classified into two main types, κ and λ, based on the amino acid sequence of their constant regions. There are three λ types and one κ, and any B cell produces only one of these. The genes for λ and κ chains and heavy chains are genetically unlinked, but, at each locus, the variable (V) genes are linked to their respective constant genes (C). In humans the location of the λ and κ light chain genes and heavy chain genes are on chromosomes 2, 22 and 14, respectively.

The κ light chain gene locus consists of a number of exons. There are two or three hundred variable regions (V$_κ$) each coding three CDRs and preceded by a leader sequence. These are followed by five J$_κ$ exons, and, finally, one constant region exon (see Figure 5). Depending on the B cell and antigen stimulus, a recombination event

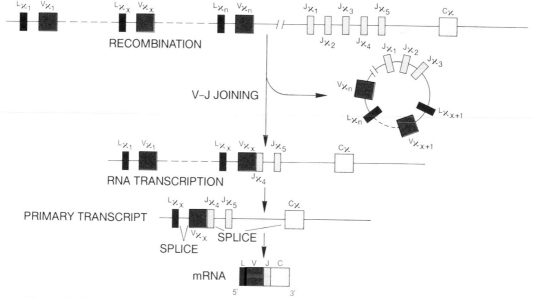

Figure 5. Genetic rearrangement of the κ light chain locus
The recombination event brings the V and J sections together and is therefore called V-J joining.
Splicing of the RNA brings about V-J-C joining, and translation of the mRNA gives rise to the
complete light chain.

occurs to bring one each of the V_κ and J_κ exons together. The join between these two
can vary (depending on the amino acid position at the point where the two exons
are joined), leading to additional diversity. This section is then transcribed to RNA
along with the constant region. RNA splicing follows bringing the entire locus together,
which can then be translated as the light chain. The λ system (λ light chain) is similar,
but with less potential for diversity. There are three constant region exons, but each
one only has one associated variable exon (Figure 6).

   The heavy chain locus is arranged similarly (see Figure 7). However, more diversity
is possible with the addition of a D region which encodes the third CDR. There are
also five types of constant exons leading to five types of antibody (IgM, IgG, IgA,
IgE and IgD). Because of the addition of a D region, two recombination events are
required now, rather than just one as with light chain gene rearrangement. The first
event brings the required D and J regions togther (D-J joining). This D-J section is
then recombined with the relevant V region to produce the V-D-J section which is
transcribed along with a specific constant region. RNA splicing and translation then
occur to produce the heavy chain. Additional diversity is achieved through somatic
mutation of certain amino acids in the variable genes. This could allow the immune
system to "fine-tune" itself, producing more and more specific antibodies after the
initial immunization event, allowing an individual to increase the antibody repertoire
from the predetermined number of genes inherited at birth.

VECTORS AND HOSTS IN THE CLONING OF ANTIBODY GENES

   To accommodate the cloning of antibody genes, a number of vector systems have
been developed, of which the most common are the pSV2 plasmids. These plasmids

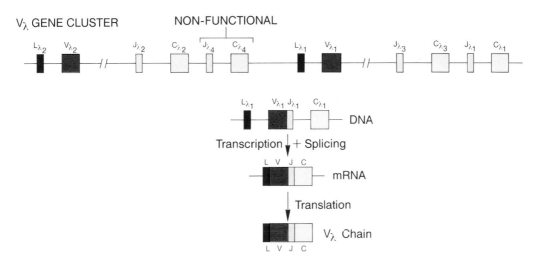

Figure 6. The V$_\lambda$ gene cluster
The recombination sequence here is the same as that of the κ light chain locus. However, the λ locus is more restricted in its diversity due to the fact that there are only three functional variable introns.

contain an origin of replication and a selection marker for bacteria. They also encode a dominant selection marker for eukaryotic cells, notably xanthine guanine phosphoribosyltransferase (gpt) which allows cells to grow in the presence of mycophenolic acid, an inhibitor of purine synthesis (see Figure 8). The choice of host used to express antibody genes is also important, and is dependent on a number of factors. Expression of functional antibodies in bacteria is difficult due to problems in folding, glycosylation[5,6] and secretion into the periplasmic space[7]. However, several successful attempts

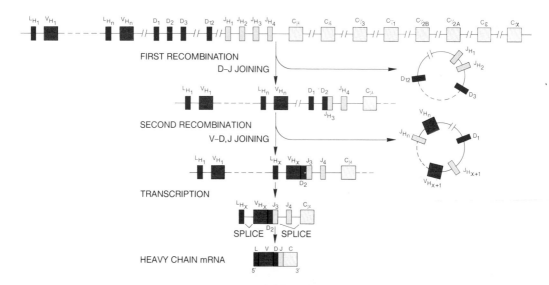

Figure 7. Genetic rearrangement of heavy chain DNA
Because of addition of a D region, two recombination events are required for the heavy chain locus, the first to bring the D and J regions together, and the second for V-D-J joining.

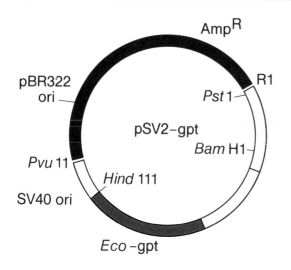

Figure 8. Diagram of the pSV2-gpt plasmid

have been made[8,9]. For example, the production of a Fab fragment which binds specifically to human carcinoma by Better *et al.*[7] who succeeded in expressing the Fab fragment in *Escherichia coli* and in getting it secreted into the medium. They used a vector designed to contain an inducible promoter from *Salmonella typhimurium* and a leader sequence derived from *Erwinia carotovora*. The advantages of using *E. coli* were the ease of handling cultures, the simple purification of the Fab fragment from culture supernatant and screening of a large number of cultures is also simplified.

Horwitz and co-workers have achieved secretion of an active chimaeric Fab fragment from yeast[10] (Figure 9b). The use of yeast overcomes some of the problems of incorrect folding and disulphide linkage formation and lack of glycosylation sometimes encountered when using bacteria. Also, yeast fermentation technology has been developed to the extent of making yeast an attractive method for expressing foreign proteins. However, there have been a number of problems such as poor antibody specificity and no antigen binding by the secreted antibody.

A baculovirus system has been used by Hasemann to express antibody fragments in insect cells[11]. These cells are eukaryotic and so correct expression, folding and secretion of antibodies can be achieved. However, insect cells are not as easily cultured as bacteria or yeast.

Hiatt and co-workers have used trangenic plants as an expression system for antibodies[12]. This has all the advantages of a eukaryotic system. Functional antibodies were expressed very efficiently in tobacco. However, plant systems have a number of drawbacks such as the limited number of techniques available to transfect cells. Furthermore, cultures are slow to grow, production levels are not as high as can be achieved with bacterial or yeast systems, and purification of the expressed protein can be difficult.

Myeloma cells are an obvious choice for use as host cells since they are capable of high level expression of heavy and light chain genes and can glycosylate, assemble and secrete functional antibody molecules. These have been used successfully on a number of occasions[13] in conjunction with several transfection techniques, notably calcium phosphate precipitation and protoplast fusion. However, transfection efficiencies using these techniques with myeloma cells are generally low (1×10^{-4}

to 1×10^{-5} transfected cells). A technique involving subjection of the cells to controlled electrical pulses, called electroporation, is now being used with myeloma cells[14].

HUMAN MONOCLONAL ANTIBODIES AND THE NEED FOR ANTIBODY ENGINEERING

Antibodies may be biologically produced by fusing antigen-sensitized spleen cells with mouse myeloma cells. The resultant antibodies are mouse in origin and this can lead to a number of problems when they are used for therapy or imaging in humans, since they can elicit an anti-mouse antibody immune response in the patient. However, it is very difficult to produce human monoclonal antibodies by conventional techniques for a number of reasons, which include difficulties with immunization (volunteers for immunization are not freely available!), the lack of good human myeloma fusion partners, the low secretion rates achieved and the instability of many human hybridomas. To date, progress has been made in the development of useful *in vitro* methods of immunization. During *in vitro* immunization the antigen of interest is added to a preparation consisting of virgin (unsensitized) human B lymphocytes, T helper cells, macrophages, and a variety of growth and stimulatory factors, including interleukins, and the mixture is left for a few days. The B cells may be derived from tonsils. B cells secreting antibody which binds to the added antigen are thus produced. Mouse myeloma cells have been used as fusion partners[15] for these B cells. The resultant heterohybridomas have good growth characteristics, but in many cases se-

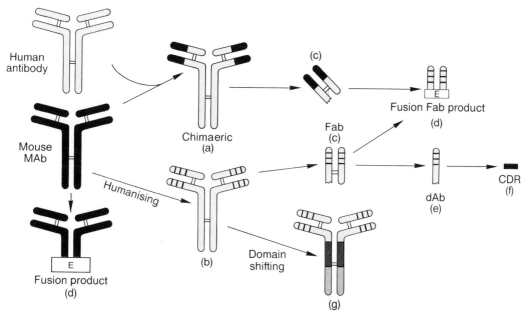

Figure 9. Antibody constructs possible by using genetic engineering techniques
Genetic engineering techniques can be used to combine human (red) and mouse (black) antibody segments to produce various constructs, as shown. (a) Chimaeric antibody (Ab) with mouse variable region and human Fc. (b) Humanised antibody with mouse CDRs. (c) Fab from humanised antibody. (d) Enzyme fused to Fab of humanised antibody. (e) Single domain humanised antibody. (f) Immunogenic protein fragment derived from CDR. (g) Humanised antibody with various Fc domains.

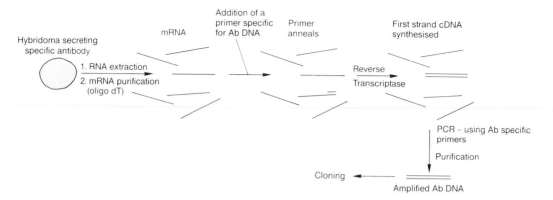

Figure 10. Isolation of antibody DNA from hybridomas

cretion of human antibody has ceased after a short time due to the preferential loss of human chromosomes. Human lymphocytes may also be immortalized by Epstein–Barr virus, and other methods of transformation, but there are problems such as low secretion levels and potential problems for therapy associated with the presence of viral particles.

A promising application of genetic engineering is the humanizing of mouse mono-clonal antibodies[5,6,16]. Advances in techniques such as the polymerase chain reaction (PCR), site-directed mutagenesis, sequencing and cloning methods have allowed pro-duction of chimaeric antibodies consisting of the variable region of a mouse antibody with all other sections derived from human sources (Figure 9). Such antibodies should reduce the possibility of the non-specific anti-mouse response in humans, but since the variable region is of mouse origin it may still elicit an immune response.

One interesting modification of the above technique is the isolation of DNA specific for a required antibody from hybridoma cells (Figure 10). Total RNA is isolated, and mRNA can be purified from this using oligo(dT) columns. The mRNA is then used as a template for the synthesis of the first strand of cDNA by way of reverse tran-scriptase. This mRNA–cDNA hybrid is then used as the template for PCR. By using probes specific for the conserved sequences in antibodies, only the required DNA will be amplified and, subsequently, purified by column chromatography or electro-phoresis. In this way, the specific antibody DNA can be isolated from hybridoma cells.

An alternative approach, as used by Lerner *et al.*, is to use spleen cells as the starting material. This would result in DNA coding for a wide range of antibodies. This can be cloned, using one of the expression systems described earlier, and an antibody library produced, which can be screened for production of antibody of the required specificity.

Research by Winter and colleagues[16] has also led to the production of humanized antibodies where only the CDR regions are of mouse origin (Figure 5b). Such reshaped antibodies would be more human-like and, therefore, less immunogenic. Humanized antibodies have been used to a limited degree for imaging cancer and other diseases *in vivo* but it may still be too early to be able to evaluate critically the long-term potential of this approach. The expertise developed may be exploited so that completely new antibodies or fragments[17] may be produced with higher affinities and with normal, or with specifically tailored, constant region functions (Figure 9g).

It should also be possible to manipulate sections of the antibody other than the variable and constant regions in order to produce improved specificity. The hinge region is very important in this respect, both in terms of its distance from the di-sulphide bonds[6] and in terms of the class of heavy chain from which it originates. Molecular biology allows manipulation of these parameters to produce designer anti-bodies[5].

Fusion antibody proteins have been successfully produced by Neuberger[18], by re-placing the Fc portion of antibody genes with such functions as nucleases (Figure 9b). In this study it was shown that these fusion antibodies could be used successfully in enzyme-linked immunoadsorbent assays for the detection of antigen, by testing for nuclease activity. Such fusion proteins have other potential uses in chemotherapy, cancer treatment, tumour imaging, etc.

When using antibodies for therapy or imaging, problems can arise with poor tissue/tumour penetration due to antibody size. Genetic engineering allows the pro-duction of any section of an antibody, and, provided that it remains functional, this solves the problem of size (and non-specific binding associated with the Fc portion). In general the approach is now to seek to produce the smallest molecular derivative of the antibody that can retain acceptable antigen binding reactivity and to test its applicability in diagnostic systems *in vitro* and *in vivo* and, eventually, in therapy *in vivo*[17,19,20]. For such approaches functions of the antibody associated with much of the constant regions may be of little significance. For example, Fab fragment produc-tion is easily achieved (Figure 9c) and this may be very useful since some of the constant region functions are conserved. Winter *et al.*[21,22] have shown that much of the specific antigen-binding capability resides in the variable region of the heavy

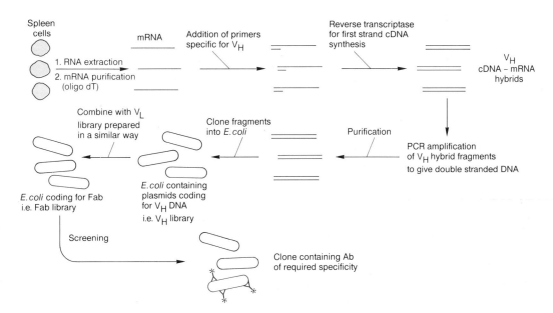

Figure 11. Scheme for producing an Fab library

chain, and have subsequently produced single-domain antibodies, consisting only of the variable heavy chain region (Figure 9e), while Williams[17] has produced biologically active peptides based only on the CDR sections of antibodies (Figure 9f).

Another interesting development is the production of antibody cDNA libraries by Lerner and co-workers[23] (Figure 11). These libraries could contain many new antibodies with improved or even completely new specificities. By combining heavy and light chain libraries prepared separately, new combinations of heavy and light chains could result, giving rise to antibodies of previously unknown specificities which could easily be screened. This would be useful in the case of catalytic antibodies where an antigen (the reaction transition state) is unavailable for immunization. This novel approach may offer great potential for antibody production. In addition, it is possible that these smaller antibody-derived units may have greater stability and, thus, have great potential for use in biosensors.

POLYMERASE CHAIN REACTION (PCR)

PCR is a technique which is becoming increasingly more useful in this area, and is now being used by a number of research groups to study various aspects of antibody engineering and so deserves a special mention. This technique requires a probe with a certain degree of homology with the target gene. However, this may not be a problem when using antibody genes since these contain many known sequences common to all antibodies, even within the frameworks of the variable genes. PCR has many benefits such as being fast, reproducible, and allows work with genes of very low copy number, and, with the development of novel techniques, it has many applications in DNA extraction, cloning, sequencing and in recombination circle

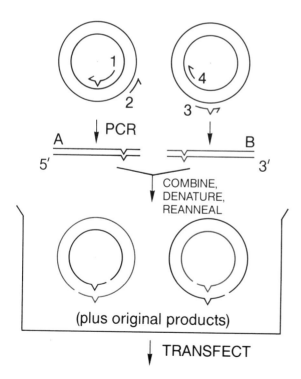

Figure 12. Recombination circle polymerase chain reaction
Two separate PCR amplifications are carried out on the same starting plasmid, giving fragments as in A and B. These have different break points due to the positioning of primers during the PCR. The primers are also designed with a base mismatch for mutagenesis. If these fragments are denatured, mixed and recombined, a fragment will be produced with protruding complementary ends, thus allowing recircularization, producing a mutated plasmid for cloning.

PCR[21,22,24], which is a system used to generate cohesive ends in a DNA fragment without the use of enzymic manipulations (Figure 12).

BIFUNCTIONAL ANTIBODIES

Bifunctional antibodies have been shown to have potential for use in immuno-chemistry, diagnostic kits, tumour imaging, therapy and sensors[2]. To date, bifunctional antibodies have been produced either by chemical linkage, or, biologically, by fusion of cells secreting antibodies of the desired specificities (Figures 13 and 14). Problems such as the large amounts of antibody required for chemical production, and the production of inactive antibody associations from quadromas (a fusion of two hybrid-omas), have meant that genetic engineering may be used to provide strategies for improving production and purification of bifunctional antibodies. It may also facilitate the production of bifunctional molecules made up predominantly from the antigen-binding portions of the antibodies. Once again, such molecules may give improved performance compared to the conventional whole antibody derivatives. We have re-cently envisaged the development of a bispecific single domain antibody[4]. Such a unit would consist of the heavy chain variable regions, which contain significant antigen binding capacity, from two different antibodies linked together. With the app-lication of antibody engineering, tri- and multi-functional antibodies with a range of specificities and capabilities are possible.

Figure 13. Chemical production of bifunctional antibodies

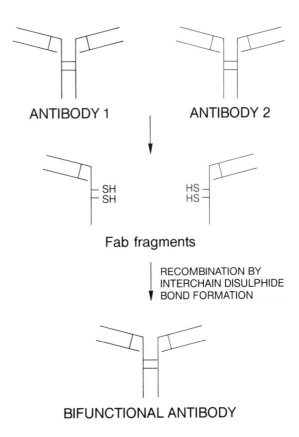

ANTIBODY 1   ANTIBODY 2

SH   HS
SH   HS

Fab fragments

RECOMBINATION BY
INTERCHAIN DISULPHIDE
BOND FORMATION

BIFUNCTIONAL ANTIBODY

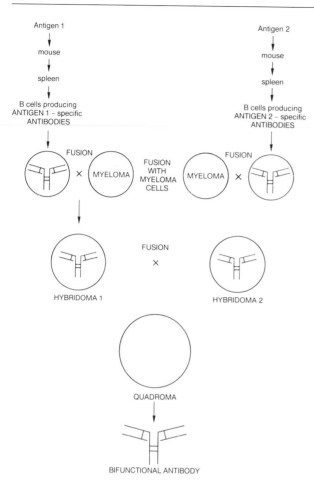

Figure 14. Biological production of bifunctional antibodies

CATALYTIC ANTIBODIES

Enzymes play a key role in catalysing many reactions. They bind to and stabilize the transition state of the reaction to be catalysed in preference to the resting or ground state. But in many cases, the transition state of a reaction is so short-lived that it is unavailable for immunization. Antibodies with such specificity for this transition state can be produced by immunizing with transition state analogues (Figure 15). Such antibodies have been shown to have significant catalytic activity for a range of reactions, including amide bond breakage[25]. Building on this, antibodies have been produced with additional sites for metals and/or cofactor binding. For example, Lerner and Benkovic have introduced a metal co-ordination site into the light chain adjacent to the antigen-binding site by using site-directed mutagenesis. It is also feasible that the catalytic activities of antibodies may be improved by protein engineering at the active site. Other approaches envisaged include the introduction of genes of antibodies with limited catalytic activities into yeast and, by using selection based on genetic complementation, it may be possible to evolve the antibody, thus producing far more efficient catalytic capability.

LARGE SCALE PRODUCTION OF ANTIBODIES

At present the major production methods for monoclonal antibodies involve the use of animals for ascitic fluid generation and large scale cell culture. The production of ascitic fluid gives good yields but there are difficulties with the use of large number of animals, the high labour costs, the presence of contaminating endogenous antibody, and possibly viruses, and ethical considerations. While large scale cell culture is technologically advanced, the costs are high, mainly due to media requirements. This has resulted in research aimed at the generation of antibodies by genetic means using a range of systems including yeast, bacteria, and baculovirus expression systems and plants[10-13,24] as previously discussed. While these technologies are still to be optimized they offer exciting new methods for antibody production on a large scale.

FUTURE DIRECTIONS

Humanizing rodent antibodies is an important development which can be achieved through genetic engineering and it has consequences for human therapy and tumour imaging[26]. Future developments in the application of genetic engineering to human antibody production may involve the initial *in vitro* immunization of human cells with antigen followed by extraction, manipulation and expression of the antibody-coding DNA from the immunized cells, leading to new human antibodies of the required specificity. The development of efficient and well-characterized expression systems will be vital for the success of this approach.

The production of more specific and improved antibodies, antibody segments and antibody hybrid proteins is also important[26]. Fusion proteins of antibodies and enzymes or toxins have uses in areas such as diagnostic assays and therapy. It is

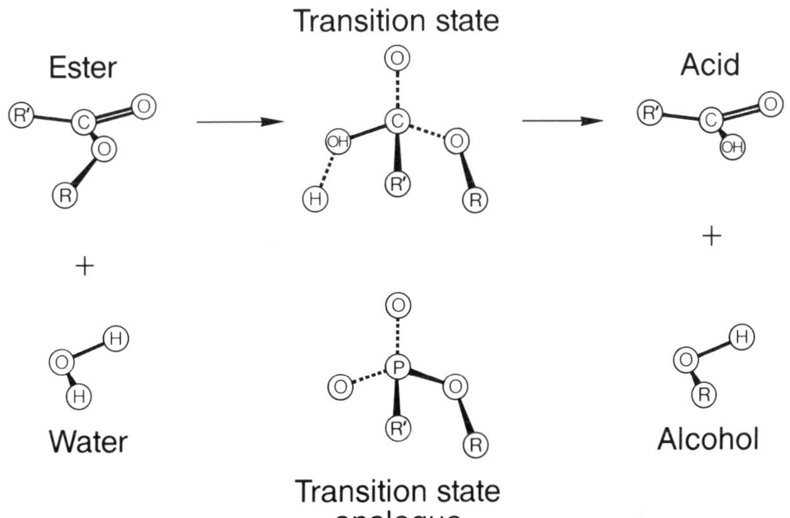

Figure 15. Ester hydrolysis
The transition state is very unstable. However, by replacing the central carbon atom with phosphorus a stable analogue can be produced, mimicking the shape and charge of the transition state.

clear that much work is required in the area of *in vitro* and *in vivo* testing to ensure that the molecules are useful, efficient and safe. It may be possible to design multimeric antibody-like proteins to carry out complex functions such as chemical chain reactions. With the development of antibody libraries, previously unknown antibodies with required specificities are easily detectable and these may be important in such areas as catalytic antibody production where no antigen is available. Antibody engineering already has been, and will continue to be, a very useful technique for studying the structure of antibodies, and elucidating how structure and function are related.

Finally, we should not forget that an antibody is itself an engineering masterpiece combining the potential for an incredible number of different antigen-binding specificities, an array of cell-binding capacities and the ability to bind complement. The fact that we can manipulate it is as much a credit to its own capability as to our ingenuity.

SUMMARY

- We can now isolate and manipulate antibody genes.
- Mouse antibodies can be humanized, resulting in chimaeric or reshaped antibodies.
- Antibody engineering is useful in large scale production of antibodies, in production of active antibody fragments, bifunctional, single-domain and catalytic antibodies, and has lead to the production of novel expression systems useful in many other areas. It allows production of new antibody conjugates, e.g. antibody–toxin or antibody–enzyme linked proteins.
- Engineered antibodies have many potential applications e.g. imaging, therapy and biosensors.

REFERENCES

1. Roitt, I., Brostoff, J. & Male, D. (1985) *Immunology*, Gower Medical Publishing, London and New York
2. Chothia, C., Lesk, A.M., Tramontano, A., Levitt, M., Smith-Gill, S., Air, G., Sheriff, S., Padlan, E., Davies, D., Tulip, W., Colman, P., Spinelli, S., Alzari, P. & Pogak, R. (1989) Conformations of immunoglobulin hypervariable regions. *Nature (London)* **342**, 877–883
3. Chothia, C. & Lesk, A. (1987) Canonical structures for the hypervariable regions of immunoglobulins. *J. Mol. Biol.* **196**, 901–917
4. Nolan, O. & O'Kennedy, R. (1990) Bifunctional antibodies: concept, production and applications. *Biochim. Biophys. Acta* **1040**, 1–11
5. Tan, L.K. & Morrison, L. (1988) Antibody structure and antibody engineering. *Adv. Drug Delivery Rev.* **2**, 129–142
6. Morrison, S.L. & Vernon, T.O. (1989) Genetically engineered antibody molecules. *Adv. Immunol.* **44**, 65–92
7. Better, M., Chang, P., Robinson, R. & Horwitz, A.H. (1988) *E. coli* secretion of an active chimeric antibody fragment. *Science* **24**, 1041–1043
8. Pluckthun, A. (1990) Antibodies from *E. coli. Nature (London)* **347**, 497–498
9. Chaudhary, V.K., Batra, J.K., Gallo, M.G., Willingham, M.C., Fitzgerald, D.G. & Pastan, I. (1990) A rapid method for cloning functional variable region antibody genes in *E. coli* as single chain immunotoxins. *Proc. Natl. Acad. Sci. U.S.A.* **87**, 1066–1070

10. Horwitz, A.H., Chang, P., Better, M., Hellstrom, K.E. & Robinson, R. (1988) Secretion of functional antibody and Fab fragment from yeast cells. *Proc. Natl. Acad. Sci. U.S.A.* **85**, 8678–8682

11. Hasemann, C.A. & Capra, J.D. (1990) High level production of a functional immunoglobulin heterodimer in a baculovirus expression system. *Proc. Natl. Acad. Sci. U.S.A.* **87**, 3942–3946

12. Hiatt, A. Cafferkey, R. & Bowdish, K. (1989) Production of antibodies in transgenic plants. *Nature (London)* **342**, 76–78

13. Vernon, T. O., Morrison, S.L., Herzenber, L.A. & Berg, P. (1983) Immunoglobulin gene expression from transformed lymphoid cells. *Proc. Natl. Acad. Sci. U.S.A.* **80**, 825–829

14. Potter, H. Weir, L. & Leder, P. (1984) Enhancer-dependent expression of human κ immunoglobulin genes introduced into mouse free B lymphocytes by electroporation. *Proc. Natl. Acad. Sci. U.S.A.* **81**, 7161–7165

15. Carroll, K., Prosser, E. & O'Kennedy, R. (1989) Parameters involved in the *in vitro* immunization of tonsillar lymphocytes: effects of rIL2 and muramyl dipeptide. *Hybridoma* **9**, 81–89

16. Riechmann, L., Clark, M., Waldmann, H. & Winter, G. (1988) Reshaping human antibodies for therapy. *Nature (London)* **332**, 323–327

17. Williams, W.V., Moss, D.A., Keiber-Emmons, T., Cohen, J.A., Myers, J.N., Weiner, D.B. & Greene, M.I. (1989) Development of biologically active peptides based on antibody structure. *Proc. Natl. Acad. Sci. U.S.A.* **86**, 5537–5541

18. Neuberger, M.S. & Fox, R.O. (1984) Recombinant antibodies possessing novel effector functions. *Nature (London)* **312**, 604–608

19. Rodwell, J.D. (1989) Engineering monoclonal antibodies. *Nature (London)* **342**, 99–100

20. McCafferty, J., Griffiths, A.D., Winter, G. & Chiswell, D.J. (1990) Phage antibodies: filamentous phage displaying antibody variable domains. *Nature (London)* **348**, 552–554

21. Orlandi, R., Gussow, D., Griffiths, A.D., Jones, P.T. & Winter, G. (1989) Cloning immunoglobulin variable domains for expression by the polymerase chain reaction. *Proc. Natl. Acad. Sci. U.S.A.* **86**, 3833–3837

22. Ward, E.S., Gussow, D.H., Jones, P.T. & Winter, G. (1989) Binding activities of a repertoire of single immunoglobulin variable domains secreted from *E. coli*. *Nature (London)* **341**, 544–546

23. Huse, W.D., Sastry, L., Iverson, S.A., Kang, A.S., Alting-Mess, M., Burton, D.R., Benkovic, S.J. & Lerner, R.A. (1989) Generation of a large combinatorial library of immunoglobulin repertoire in phage. *Science* **246**, 1275–1280

24. Jones, D.H., Sakamoto, K., Vorce, R.L. & Howard, B.H. (1990) DNA mutagenesis and recombination. *Nature (London)*, **344**, 793–794

25. Lerner, R.A. & Tramontano, A. (1988) Catalytic antibodies. *Sci. Am.* **258(3)**, 58–70

26. Winter, G. & Milstein, C. (1991) Man-made antibodies. *Nature (London)* **349**, 293–299

7

Ecstasy: towards an understanding of the biochemical basis of the actions of MDMA

Marcus Rattray

UMDS Division of Biochemistry, University of London, St Thomas's Campus, London SE1 7EH, U.K.

INTRODUCTION

Illegal use of the mood-altering drug methylenedioxymethamphetamine (MDMA), commonly known as "ecstasy" or "E", is now firmly established in Europe and the U.S.A. MDMA abuse is of concern, particularly because animal studies have indicated that MDMA can induce long-term neuronal damage. In this article I review some of the complex biochemical actions of MDMA and discuss how these may relate to the psychopharmacological and neurotoxic effects of this drug.

USES AND ABUSES OF MDMA

MDMA was patented in 1914 for use as an appetite suppressant (anorectic) but never marketed. In the 1970s MDMA re-emerged on a small scale as a tool to facilitate psychotherapy because of its effects on mood: MDMA was claimed to break down the barriers between psychiatrist and patient. From these medically-controlled uses, MDMA rapidly became popular as a recreational drug. The drug was made illegal in the U.S.A. and U.K. in the mid-1980s because of concerns about its abuse potential, the absence of therapeutic applications and the possibility that MDMA may cause

Name	Structure	Medical use
Amphetamine (speed, benzedrine)		Narcolepsy, hyperactivity in children, anorectic
Methamphetamine (speed)		As amphetamine
3,4-Methylenedioxyamphetamine (MDA, love drug, ice)		None
3,4-Methylenedioxymethamphetamine (MDMA, Ecstasy, E)		None
Fenfluramine		Anorectic
Fluoxetine		Antidepressant
5-Hydroxytryptamine (serotonin, 5-HT)		None

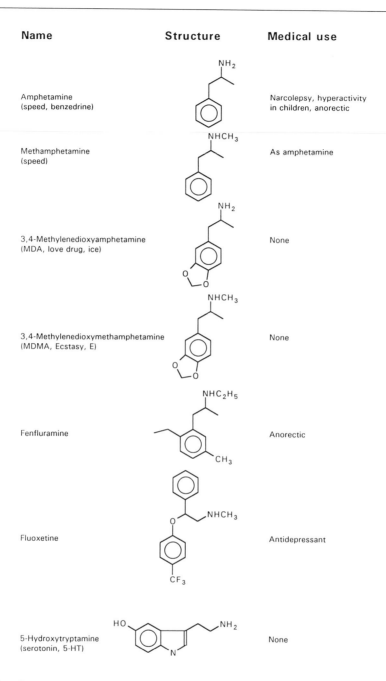

Figure 1. Structures

neurotoxicity. Despite its illegality, its use has continued to grow. There are no approved clinical uses for hallucinogenic amphetamines such as MDMA.

The popularity of MDMA can be ascribed to its psychotropic (mood-altering) effects. MDMA is a phenylisopropylamine, closely related to its parent compound, amphetamine (Figure 1). MDMA causes a curious mixture of amphetamine-like stimulant effects as well as hallucinogenic effects. Users of MDMA experience a 5–6 hour "high" which produces increased activity, mood elevation and alterations in perception.

In most cases there do not appear to be any long-term consequences of MDMA abuse. For some individuals, MDMA can cause a severe acute reaction, including hyperthermia, alterations in cardiovascular function, respiratory distress, rhabdomyolysis and intravascular coagulation, which may result in death[1].

ACTIONS OF MDMA ON 5-HYDROXYTRYPTAMINE NEURONES

Current evidence suggests that the primary action of MDMA is on nerve terminals of neurones that synthesize and release the amine neurotransmitter 5-hydroxy-

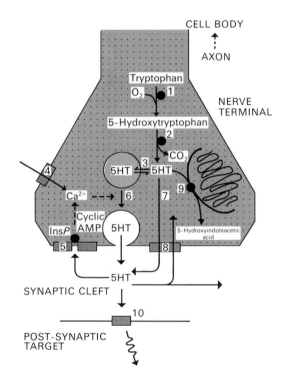

Figure 2. Model of a 5-hydroxytryptamine-containing nerve terminal
Key: 1, tryptophan hydroxylase, the rate-limiting enzyme of 5HT synthesis; 2, aromatic amino acid decarboxylase, which converts 5-hydroxytryptophan to 5-hydroxytryptamine (5HT); 3, the vesicular 5HT transporter, ATP-dependent, transports 5HT into storage vesicles; 4, a voltage-sensitive Ca^{2+} channel, which opens in response to depolarization, causes 5HT release from vesicles; 5, presynaptic 5HT receptors: various subtypes may be coupled, via G proteins, to adenylate cyclase or phospholipase C, thus producing autoregulation of 5HT release; 6, calcium-dependent 5HT release from vesicles, probably mediated by actions of protein kinase C on vesicular proteins such as synapsin; 7, calcium-independent 5HT release, efflux of cytoplasmic stores of 5HT through the 5HT transporter; 8, the synaptic membrane 5HT transporter, which uses a sodium ion gradient to take up released 5HT from the synaptic cleft; 9, monoamine oxidase, a mitochondrial-bound enzyme that metabolizes 5HT to 5-hydroxyindoleacetic acid, its major endogenous metabolite; 10, post-synaptic 5HT receptors: activation by released 5HT causes depolarization or hyperpolarization of target neurones and leads to the complex psychopharmacological actions of 5HT.

tryptamine (serotonin; 5HT) (Figure 2). The structure of 5HT is shown in Figure 1. The effects of MDMA are stereoselective, the + isomer being the most potent, although for most studies the racemic mixture has been used. MDMA is an indirect 5HT agonist, causing a release of 5HT, inhibition of its synthesis and a block of its reuptake into nerve terminals. MDMA may also modulate the effects of endogenous 5HT that is released, since it binds to several neurotransmitter receptors, including 5HT receptors.

DEPLETION OF 5-HYDROXYTRYPTAMINE *IN VIVO* BY MDMA

In the rat a single injection of MDMA (10-40 mg/kg) causes a significant reduction in the tissue content of 5HT and its major metabolites in several brain regions that receive major 5HT innervation from midbrain 5HT cell groups[2-4]. There are subtle differences in the results obtained by different laboratories but all agree that 5HT or metabolite levels drop to at least 30% of controls, although not all brain regions are affected equally[4].

After a single dose of MDMA, 5HT depletion is rapid, occurring within 1–3 hours of drug administration. Levels remain low for 6–18 hours and recover within 24 hours[2-4]. The time course of 5HT depletion corresponds well to the time course of the behavioural effects of MDMA. Thus it is likely that the psychotropic effects of MDMA can be ascribed to the post- and pre-synaptic effects of released 5HT.

RELEASE OF 5-HYDROXYTRYPTAMINE BY MDMA

Studies *in vitro* using brain slices preloaded with [^{3}H]5HT have shown that micromolar concentrations of MDMA induce 5HT release[5,6]. It has been proposed that the MDMA taken up by nerve terminals causes the displacement of 5HT from cytoplasmic binding sites, leading to 5HT efflux through the synaptic membrane 5HT transporter. Consistent with this hypothesis, 5HT release appears to be calcium-independent[6]. For 5HT, this is taken as evidence that the neurotransmitter released is derived from cytoplasmic stores rather than from 5HT stored in synaptic vesicles[7].

5-HYDROXYTRYPTAMINE TRANSPORTERS

MDMA has been found to have submicromolar binding affinity to a protein consistent with a high-affinity, synaptic membrane 5HT transporter (serotonin uptake site) that co-transports Cl^- and Na^+ down their concentration gradients[8]. This transporter is one member of a class of proteins that can transport 5HT into nerve terminals, synaptic vesicles or platelets. Since studies *in vitro* have shown that MDMA is capable of inhibiting the uptake of [^{3}H]5HT into nerve terminals[2], it seems that MDMA has antagonist actions at this site *in vivo*. A blockade of 5HT reuptake into the nerve terminal would prolong the pre- and post-synaptic actions of released 5HT by effectively increasing the concentration of 5HT in the synaptic cleft.

Despite its high affinity for a 5HT transporter, it is not clear whether MDMA is itself transported into neurones. It has been suggested that amphetamines such as MDMA are taken into neurones by a passive process that does not involve a specific transport protein[9].

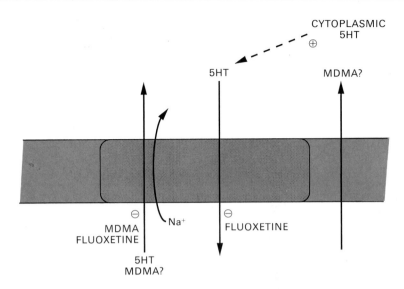

Figure 3. Actions of MDMA involving the 5HT transporter
MDMA blocks 5HT uptake, but causes 5HT release, apparently through the transporter. It is not clear whether MDMA enters the nerve terminal, but if so, it may displace 5HT from cytoplasmic binding sites. Fluoxetine blocks 5HT uptake and does not cause 5HT release.

BLOCKADE OF 5-HYDROXYTRYPTAMINE RELEASE BY FLUOXETINE

Drugs that are known to block 5HT uptake into nerve terminals but do not themselves release 5HT (for example, fluoxetine; see Figure 1) are found to inhibit release of 5HT induced by MDMA *in vitro* and 5HT depletion *in vivo*[2,5].

These observations suggest several hypotheses, outlined in Figure 3. First, MDMA may be taken up into neurones by the transporter, or by a passive mechanism, and once inside the cell may displace 5HT from cytoplasmic stores; this 5HT is then released through a synaptic membrane 5HT transporter. In this case, fluoxetine could block the uptake of MDMA into nerve terminals. A second possible hypothesis is that MDMA is not taken up into nerve terminals at all, but causes 5HT release by binding to the synaptic membrane 5HT transporter. In this case, fluoxetine might block the binding of MDMA to the transporter, and the subsequent intracellular events caused by MDMA binding that induce release of 5HT.

To distinguish fully between these hypotheses requires a detailed biochemical knowledge of 5HT transporters that is not yet available. Whereas some drugs, such as MDMA, that bind to the protein apparently block 5HT import and induce 5HT efflux, other drugs such as fluoxetine appear to block import of 5HT without causing 5HT efflux. The nature of the biochemical events that underlie the relationship between drug binding and 5HT release or uptake are only beginning to be understood[10].

INTERACTIONS OF MDMA WITH NEUROTRANSMITTER RECEPTORS

Although MDMA releases 5HT, thus producing an indirect activation of a variety of different 5HT receptor subtypes, some of the effects of MDMA may be mediated by direct interaction of MDMA with neurotransmitter receptors. Apart from its high affinity for the 5HT transporter, MDMA has binding affinity in the micromolar range for a number of neurotransmitter receptors: the $5HT_2$ receptor, the M_1 muscarinic acetylcholine receptor, the α_2 adrenergic receptor and the histamine H1 receptor[8].

There are correlations between certain of these receptor interactions and the clinical manifestations of MDMA intoxication. The interaction of MDMA with α_2 adrenergic receptors may be physiologically relevant for some of the cardiovascular effects of MDMA in increasing cardiac output and hypertension and inducing arrhythmias.

The $5HT_2$ receptor subtype is considered to be of major importance for the actions of classical hallucinogenic drugs such as LSD. It is tempting therefore to speculate that MDMA binding to this site may modulate the effects of released 5HT and thus produce the hallucinogenic effects experienced during MDMA use.

DOES MDMA INTERACT WITH RECEPTORS *IN VIVO*?

From binding studies it is not known whether MDMA has agonist- or antagonist-like actions upon receptor proteins or, indeed, whether the brain concentration of MDMA after drug administration is sufficient to allow significant MDMA binding to receptors *in vivo*. The second question has been partly addressed, in the rat, by studying the tissue distribution of the closely related hallucinogenic amphetamine, MDA (Figure 1). At 45 minutes after a single injection of a behaviourally active dose (20 mg/kg) the brain concentration of MDA was found to be 165 μM[9]. The half-life of brain MDA was approximately 2 hours. These results imply that after peripheral administration brain concentrations of MDMA may be sufficient to occupy neurotransmitter receptors that have micromolar affinity for MDMA.

EFFECT OF MDMA ON DOPAMINE SYSTEMS

Although amphetamine and metamphetamine are potent dopamine releasing agents *in vivo* and *in vitro*, MDMA is very selective in its depletion of 5HT. *In vitro*, MDMA has been demonstrated to block uptake of dopamine and noradrenaline[6] although it has a low binding affinity for a dopamine transporter protein[8]. There is little evidence to suggest that MDMA can directly cause release of dopamine *in vivo* in most brain regions.

Several groups have found an effect of MDMA upon a group of dopamine neurones that project from the substantia nigra to the striatum. The exact effect of MDMA *in vivo* on nigrostriatal dopamine neurones is not entirely clear, but most reports now suggest that MDMA inhibits dopamine release and neuronal firing[11]. The effect of MDMA upon nigrostriatal dopaminergic neurones appears to be an indirect one mediated by local 5HT release. It is likely that this interaction mediates the stimulant effects of MDMA.

EFFECT OF MDMA ON TRYPTOPHAN HYDROXYLASE ACTIVITY

In parallel with a reduction in 5HT levels due to its release, single high doses of MDMA have been shown to produce a decrease in the activity of tryptophan hydroxylase, the rate-limiting enzyme of 5HT synthesis[3,12,13] (Figure 4). Tryptophan hydroxylase activity falls more rapidly than 5HT levels and its recovery is much slower. Although the initial effect is reversible, 2 weeks after a single injection of MDMA (20 mg/kg), cerebral cortex and hippocampal tryptophan hydroxylase activities remain significantly depressed, implying that freshly synthesized enzyme is necessary to replace the lost activity. Thus the actions of MDMA upon the pool of tryptophan hydroxylase in the nerve terminal become irreversible after a few hours and can be considered to be toxic.

The precise mechanism by which MDMA interferes with tryptophan hydroxylase activity is unknown. Recent studies, however, have shown that cerebral cortical tryptophan hydroxylase activity from MDMA-treated animals (up to 3 hours post-injection) can be recovered *in vitro* by reduction with sulphydryl reagents under anaerobic conditions[13]. This fits with the observations that tryptophan hydroxylase is an extremely labile enzyme: its sulphydryl groups are sensitive to oxidation, including oxidation by molecular oxygen, its substrate[14]. It seems likely therefore that MDMA directly or indirectly causes oxidative stress in nerve terminals that can inactivate tryptophan hydroxylase.

MDMA-INDUCED NEURODEGENERATION

After administration of four to eight doses of MDMA (2.5–20 mg/kg) over 2 or 4 days, MDMA produces axon degeneration of 5HT-containing neurones in rodents and non-human primates[15,16]. Axonal degeneration is associated with a decrease in biochemical markers of 5HT nerve terminals such as the 5HT transporter and 5HT content[17]. The biochemical events that occur to allow toxicity to progress from short-term events such as tryptophan hydroxylase inactivation to irreversible degeneration are unclear, but it seems that the manifestations of nerve terminal loss are dose- and time-dependent.

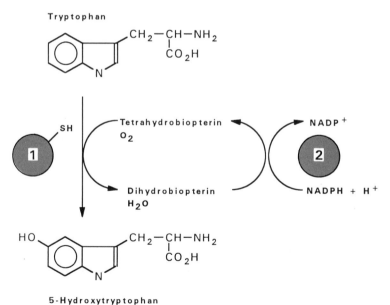

Figure 4. Tryptophan hydroxylase
Tryptophan hydroxylase (1) catalyses the oxidation of tryptophan to 5-hydroxytryptophan. The enzyme requires the cofactor tetrahydrobiopterin and its associated, NADPH-dependent, reducing system, dihydrobiopterin reductase (2). Tryptophan hydroxylase itself is extremely sensitive to oxidation.

From time points as early as 18 hours after the last drug injection, immunocyto-chemical studies have shown that there is a huge loss in cerebral cortical and striatal 5HT content accompanied by the appearance of swollen and fragmented 5HT axons. The recovery from MDMA administration is very slow — restoration of normal cortical 5HT levels does not occur until 1 year after cessation of drug administration in rats[17]. Interestingly, despite the massive morphological changes in axons, 5HT cell bodies are not apparently affected[15].

Not all 5HT neurones are equally affected by MDMA. In many brain regions such as cerebral cortex, the fine axons which probably originate from the dorsal raphe nucleus are lost, whereas beaded axons which originate in the medial raphe nucleus are selectively spared[16]. It seems unlikely that this selective neurotoxicity reflects the access of MDMA to different brain regions, since brain areas where selective cell loss occurs may have numerous projections from both dorsal and medial raphe. These findings are intriguing, and imply that biochemical differences exist between sub-groups of 5HT neurones that underlie their differential sensitivity to MDMA-induced toxicity.

MECHANISM OF MDMA-INDUCED TOXICITY

MDMA toxicity appears dependent upon calcium-independent 5HT efflux, a sy-naptic membrane 5HT transporter[12] and calcium ions[18,19] since compounds that block the transporter or prevent increases in intracellular calcium prevent toxicity. For example, administration of a 5HT uptake blocker, fluoxetine, up to 6 hours after MDMA injection is found to reverse the MDMA-induced reduction in tryptophan hydroxylase activity[12]. The mechanisms are unknown, but it is likely that MDMA induces oxidative stress in nerve terminals, which leads to inactivation of tryptophan hydroxylase and further events that lead to irreversible nerve terminal damage. These are illustrated in Figure 5.

MDMA can be metabolized to a quinoid[20] that may enter the nerve terminal and directly combine with sulphydryl groups within the active site of tryptophan hydroxylase. Alternatively, redox cycling of MDMA metabolites may generate reactive species such as free radicals that can inactivate tryptophan hydroxylase and cause damage to protein and lipid components of the nerve terminal.

MDMA binding to the 5HT transporter or a 5HT receptor may set off a chain of intracellular events that lead to tryptophan hydroxylase inactivation, for instance by increasing the intracellular calcium ion concentration[18,19]. Elevated calcium might lead to protein kinase C-mediated actions such as proteolysis, as well as to osmotic stress that could lead to degeneration.

Alternatively, MDMA may inhibit monoamine oxidase and thus allow auto-oxida-tion of 5HT or dopamine to a known toxin (e.g. 5,7-dihydroxytryptamine); however, the presence of these toxins in the brain after MDMA administration has not yet been demonstrated.

IS MDMA TOXIC TO MAN?

An important question arising from these studies is whether MDMA has neurotoxic effects in humans. The key question is whether the pattern of drug-taking in humans (for example, one or two tablets containing 95 mg of MDMA per week) produces

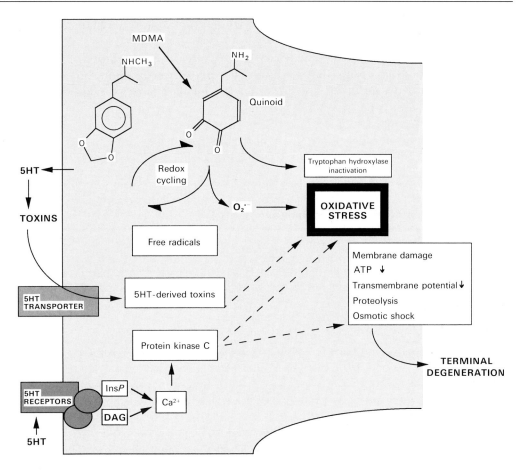

Figure 5. Some possible mechanisms of MDMA-induced toxicity
MDMA could induce toxicity by a variety of mechanisms including: metabolism to a quinoid,
generation of free radicals, production of a 5HT-derived toxin, or by increasing intracellular
calcium ion concentration. The mechanism is unknown but probably involves the induction of
oxidative stress in the terminal. This can progress to irreversible terminal degeneration.

any irreversible effects on 5HT neurones. In all the studies that have found neurodegeneration in animals, several large doses were administered over a very short
time period, so it is difficult to extrapolate from animals to humans. The route of
drug administration (oral in humans) is a significant factor[21]. Nevertheless, it is likely
that the level of consumption in man can produce brain concentrations that approach
toxic doses. At the present time there are no reports of MDMA-induced neurodegeneration in humans.

The behavioural effects of 5HT depletion in man are not known. Psychopharmacological studies suggest that the 5HT system is strongly associated with the generation
and maintenance of mood, so it may be that loss of 5HT neurones may lead to
long-term neuropsychiatric illnesses. Indeed, a recent report indicates that excessive
MDMA use can lead to long-term psychosis[22]. However, any effects of MDMA in

destroying even a small percentage of 5HT terminals in humans may not be manifested for many years.

For the neuroscientist, drugs such as MDMA offer the possibility to explore not only the biochemical actions that underlie 5HT neurotransmission, but also may be used as a tool to explore the link between the biochemistry of neurotransmission and psychological events. Since MDMA is a selective neurotoxin, it may provide a key to understand the phenomenon of selective neuronal death. An understanding of the mechanism of action of MDMA may therefore relate not only to drug abuse but to disorders of mood and neurological illnesses involving the 5HT system.

Many thanks to Monica Rattray for the figures, and apologies to the many authors of the 50 or so MDMA papers that were not included here for reasons of space. My research is funded by the Medical Research Council and the Special Trustees of St. Thomas's Hospital.

REFERENCES

1. Dowling, G.P., McDonough III, E.T. & Bost, R.O. (1987) "Eve" and "Ecstasy". A report of five deaths associated with the use of MDEA and MDMA. *J. Am. Med. Assoc.* **257**, 1615–1617

2. Schmidt, C.J. (1987) Neurotoxicity of the psychadelic amphetamine, methylenedioxymeth-amphetamine. *J. Pharmacol. Exp. Ther.* **192**, 33–41

3. Stone, D.M., Stahl, D.C., Hanson, G.R. & Gibb, J.W. (1986) The effects of 3,4-methylene dioxymethamphetamine (MDMA) and 3,4-methylenedioxyamphetamine (MDA) on monaminergic systems in the rat brain. *Eur. J. Pharmacol.* **128**, 2294–2303

4. Mokler, D.J., Robinson, S.E. & Rosencraus, J.A. (1987) (+)-3,4-Methylenedioxymeth amphetamine produces long-term reductions in brain 5-hydroxytryptamine in rats. *Eur. J. Pharmacol.* **138**, 265–268

5. Hekmatpanah, C.R. & Peroutka, S.J. (1990) 5-Hydroxytryptamine uptake blockers attenuate the 5-hydroxytryptamine releasing effect of 3,4-methylenedioxymethamphet-amine and related agents. *Eur. J. Pharmacol.* **177**, 95–98

6. Johnson, M.P., Hoffman, A.J. & Nichols, D.E. (1986) Effects of the enantiomers of MDA, MDMA and related analogues on [^{3}H]serotonin and [^{3}H] dopamine release from superfused rat brain slices. *Eur. J. Pharmacol.* **132**, 269–276

7. Fuller, R.W. & Hemrick-Luecke, S.K. (1982) Further studies on the long-term depletion of striatal dopamine in iprindole treated rats by amphetamine. *Brain Res.* **447**, 141–144

8. Battaglia, G., Brooks, B.P., Kulsakdinun, C. & De Souza, E.B. (1988) Pharmacologic profile of MDMA (3,4-methylenedioxymethamphetamine) at various brain recognition sites. *Eur. J. Pharmacol.* **149**, 159–163

9. Zaczek, R., Hurt, S., Culp, S. & De Souza, E.B. (1989) Characterisation of brain interactions with methylenedioxyamphetamine and methylenedioxymethamphetamine. *NIDA Res. Monogr.* **94**, 223–239

10. Biessen, E.A.L., Horn, A.S. & Robillard, G.T. (1991) Approaches to the purification of the 5-hydroxytryptamine reuptake system from human blood platelets. *Biochem. Soc. Trans.* **19**, 103–111

11. Gazzara, R.A., Takeda, H., Cho, A.K. & Howard, S.G. (1989) Inhibition of dopamine release by methylenedioxymethamphetamine is mediated by serotonin. *Eur. J. Pharmacol.* **168**, 209–217

12. Schmidt, C.J. & Taylor, V.L. (1987) Depression of rat brain tryptophan hydroxylase activity following the acute administration of methylenedioxymethamphetamine. *Biochem. Pharmacol.* **36**, 4095–4102

13. Stone, D.M., Johnson, M., Hanson, G.R. & Gibb, J.W. (1989) Acute inactivation of tryptophan hydroxylase by amphetamine analogs involves the oxidation of sulphydryl sites. *Eur. J. Pharmacol.* **172**, 93–97

14. Kuhn, D.M., Ruskin, B. & Lovenberg, W. (1980) Tryptophan hydroxylase: the role of oxygen, iron and sulphydryl groups as determinants of stability and catalytic activity. *J. Biol. Chem.* **255**, 4137–4143

15. O'Hearn, E., Battaglia, G., De Souza, E.B., Kuhar, M.J. & Molliver, M.E. (1988) Methylene-dioxyamphetamine (MDA) and methylenedioxymethamphetamine (MDMA) cause ablation of serotonergic axon terminals in forebrain: immunocytochemical evidence. *J. Neurosci.* **8**, 2788–2803

16. Wilson, M.A., Ricaurte, G.A. & Molliver, M.E. (1989) Distinct morphological classes of serotonergic axons in primates exhibit differential vulnerability to the psychotropic drug 3,4-methylenedioxymethamphetamine. *Neuroscience* **28**, 121–137

17. Battaglia, G., Yeh, S.Y., O'Hearn, E., Molliver, M.E., Kuhar, M.J. & De Souza, E.B. (1987) 3,4-Methylenedioxymethamphetamine and 3,4-methylenedioxyamphetamine destroy serotonin terminals in rat brain: quantification of neurodegeneration by measurement of [^{3}H]paroxetine labelled serotonin uptake sites. *J. Pharmacol. Exp. Ther.* **242**, 911–916

18. Azmitia, E.C., Murphy, R.B. & Whitaker-Azmitia, F.M. (1990) MDMA (Ecstasy) effects on cultured neurons: evidence for Ca^{2+} dependent toxicity linked to release. *Brain Res.* **510**, 97–103

19. Finnegan, K.T., Skratt, J.J., Irwin, I. & Langston, J.W. (1990) The *N*-methyl-D-aspartate (NMDA) receptor antagonist dextrorphan prevents the neurotoxic effects of 3,4-methylenedioxymethamphetamine (MDMA) in rats. *Neurosci. Lett.* **105**, 300–306

20. Hiramatsu, M., Kumagai, Y., Unger, S.E. & Cho, A.K. (1990) Metabolism of methylene-dioxymethamphetamine: formation of dihydroxymethamphetamine and a quinone identified as its glutathione adduct. *J. Pharmacol. Exp. Ther.* **254**, 521–527

21. Ricaurte, G.A. (1989) Studies of MDMA-induced neurotoxicity in nonhuman primates: a basis for evaluating long-term effects in humans. *NIDA Res. Monogr.* **94**, 306–322

22. McGuire, P. & Fahy, T. (1991) Chronic paranoid psychosis after misuse of MDMA ("ecstasy"). *Br. Med. J.* **302**, 697

8

Structure and function of ribonuclease A binding subsites

Xavier Parés, M. Victòria Nogués, Rafael de Llorens and Claudi M. Cuchillo

Departament de Bioquímica i Biologia Molecular, Facultat de Ciències and Institut de Biologia Fonamental "Vicent Villar Palasí", Universitat Autònoma de Barcelona, 08193 Bellaterra (Barcelona), Spain

INTRODUCTION

Bovine pancreatic ribonuclease A (EC 3.1.27.5) has a molecular mass of 13.7 kDa and 124 amino acid residues. It is a well defined enzyme, with known chemical and crystallographic structures. It has been used as a test protein in the study of a wide variety of chemical and physical methods applied to protein chemistry[1–3]. Ribonuclease A catalyses the hydrolysis of 3′,5′-phosphodiester linkages of single-stranded RNA at the 5′-ester bond, in a two-step reaction (Figure 1). The first step is a trans-phosphorylation reaction which yields a 2′,3′-cyclic phosphate terminus and a free 5′-OH group on the other side of the bond cleaved. The second step is the hydrolysis of the cyclic phosphodiester to give a terminal 3′-phosphate monoester. The base at the 3′-side must be a pyrimidine (uracil or cytosine). The base at the 5′-side can be either a pyrimidine or purine.

Much research work during the last two decades has unambiguously demonstrated the existence of multiple binding subsites in ribonuclease A that essentially recognize the negatively charged phosphates of RNA[4]. We shall summarize here these studies and discuss the possible role of the subsites in the binding, specificity and catalytic properties of ribonuclease A. We shall also consider whether the subsite structure of

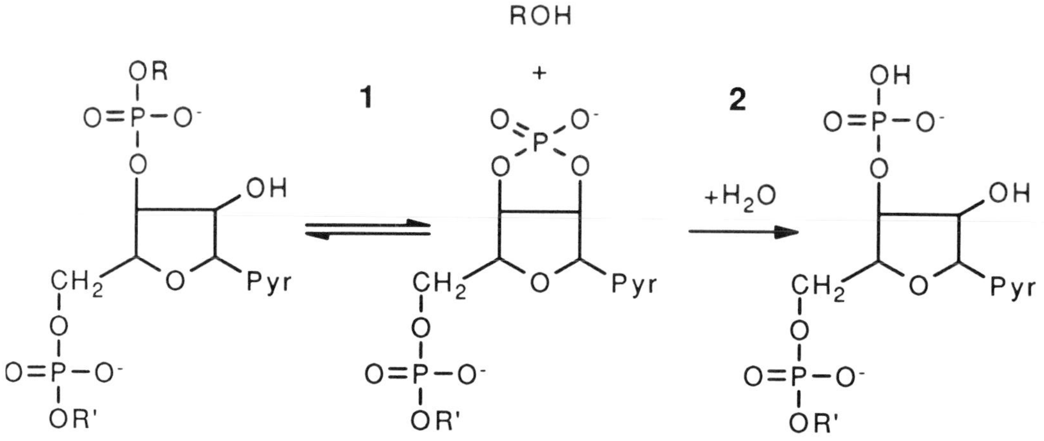

Figure 1. The two-step hydrolysis of RNA by ribonuclease A
In the transphosphorylation step a cyclic phosphate intermediate is formed which is hydrolysed in the subsequent step. The cyclization is usually much faster than the hydrolysis, so the intermediate may be readily isolated. DNA is not hydrolysed, as it lacks the 2'-hydroxyl group that is essential for this reaction. There is a strong specificity for the base B on the 3' side of the substrate to be a pyrimidine.

this enzyme can be used as a general model for the interaction of nucleic acids with nucleases and other nucleic acid binding proteins.

THE ACTIVE SITE

In the early 1970s, a structural scheme of the active site was deduced from the important crystallographic studies by Richards, Wyckoff and coworkers[1]. These authors analysed complexes of ribonuclease S (an active derivative produced by proteolytic cleavage between Ala-20 and Ser-21) and nucleotides (2'-CMP, 3'-CMP, 5'-AMP) or dinucleotide substrate analogues such as UpcA [UpcA corresponds to

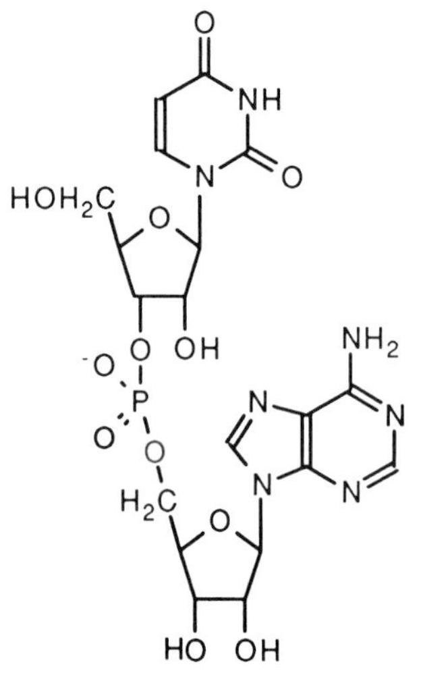

Figure 2. Structure of UpA
In the analogue UpcA the 5'-oxygen atom (in red) is replaced by a -CH$_2$- group.

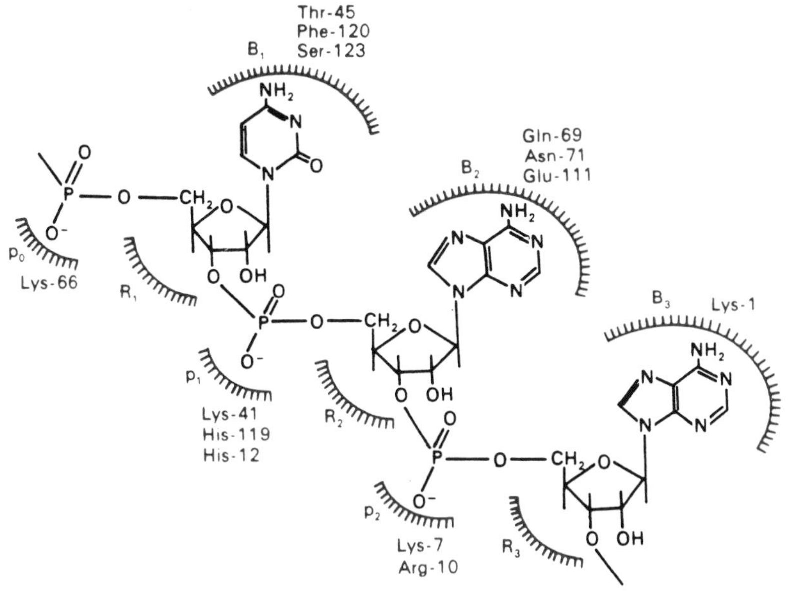

Figure 3. Schematic diagram of the active centre cleft in the ribonuclease A–substrate complex

B, R, and p indicate binding subsites for base, ribose and phosphate, respectively. B_1 is specific for pyrimidines and B_2 "prefers" purines. 3'-Pyrimidine mononucleotides bind to $B_1R_1p_1$. 5'-Purine mononucleotides bind to $B_2R_2p_1$. 3'-AMP binds to $B_2R_2p_2$. The phosphate group of the phosphodiester bond hydrolysed by the enzyme binds to p_1. The residues known to be involved in each site are indicated[11].

the usual dinucleoside phosphate UpA (Figure 2), except that the 5'-oxygen atom of the ribose attached to the adenine has been replaced with a methylene group; the phosphorus–carbon bond thus formed cannot be cleaved by ribonuclease]. B_1, R_1, p_1, R_2 and B_2 indicate the relative positions of the bases, riboses and phosphate of UpcA (Figure 3). The 2'- and 3'-pyrimidine nucleotides occupy B_1, R_1 and p_1 predominantly. 5'-AMP binds at B_2, R_2 and p_1. The His-119, His-12 and Lys-41 side chains interact with the phosphate group of the ligands, and His-12 with the 2'-hydroxyl group of the dinucleotide. These studies also revealed the existence of hydrogen bonds of both the ribose and pyrimidine base with Thr-45, Asn-44, and possibly Ser-123, as well as van der Waals contacts with the side chain of Phe-120 (B_1R_1 sites). The adenine moiety of UpcA interacts with Gln-111, Glu-69, Asn-71 and Ala-109 (B_2 site).

More recently the crystal structure of ribonuclease A has been studied in depth by the groups of Wlodawer and Borkakoti. Their work demonstrated that the structures of ribonuclease A and ribonuclease S are similar except for regions near the cleavage site (residues 17–23) (for review see[5]). The protein is U-shaped, with aproximate dimensions of 3.5 nm × 4.5 nm × 3.1 nm. A pronounced cleft forms the binding site for substrates. A further refinement of the ribonuclease structure was performed by neutron diffraction studies. This allowed the reposition of the important side chain residues His-119, His-12 and Lys-41 in the active site. The reinvestigation of the complexes with nucleotides and analogues confirmed the general structure of $B_1R_1p_1R_2B_2$ previously discussed. Uridine vanadate (a complex prepared by mixing uridine and

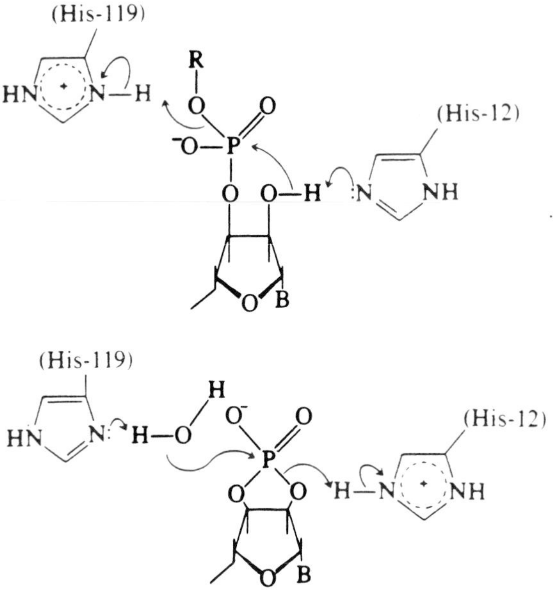

Figure 4. Currently accepted chemical mechanism for the ribonuclease A catalytic reaction
In the transphosphorylation step, His-12 acts as a general-base catalyst and His-119 acts as a general acid to protonate the leaving group. Their catalytic roles are reversed in the hydrolysis step: His-119 activates the attack of water by general-base catalysis and His-12 is the acid catalyst, protonating the leaving group.

vanadate in a 1.4:1 ratio) has been used as a putative transition state analogue. The crystallographic study of its interaction with ribonuclease A suggests that Lys-41 is involved in the stabilization of the transition state complex[5].

The components of the p_1R_1 binding sites are responsible for the currently accepted chemical mechanism of the reaction, which is explained in Figure 4. The chemistry and stereochemistry of the reaction have been convincingly determined[6]. There is an in-line mechanism that generates a pentacovalent intermediate, with the attacking nucleophile and the leaving group occupying the apical positions of the trigonal bipyramid. As previously discussed, Lys-41 would contribute to the stability of the intermediate.

Other kinetic properties can also be explained in terms of the different binding sites at the active centre. Thus, the transphosphorylation reaction of CpA proceeds with a $k_{cat.}$ of 3000 s^{-1} and K_m 1.0 mM, while $k_{cat.}$ for CpU is 27 s^{-1} and K_m 3.7 mM. The difference in rates is attributed to a separate binding site for the base at the 5'-side of the phosphodiester bond. The binding of adenosine induces a conformational change in the active centre, involving His-119 and Asp-121, which stimulates the hydrolysis of 2',3'-cyclic CMP[7].

THE p_0 BINDING SITE

A phosphate binding subsite in the 5'-side of the ribose at R_1 (see Figure 3) was first proposed to explain the binding characteristics of uridine-2'(3'),5'-bisphosphate (a mixture of the uridine 2',5'-bisphosphate and uridine 3',5'-bisphosphate isomers) with ribonuclease A and with carboxymethylribonuclease A: (a) the association constant of the nucleotide diphosphate is 14 times stronger than that of 2'(3')-UMP (mixture of the 2'-UMP and 3'-UMP isomers), and (b) 2'(3')-UMP virtually does not bind to the p_1-site blocked enzyme, carboxymethylhistidine[119]-ribonuclease A, while uridine 2'(3'),5'-bisphosphate still binds to the modified enzyme with strong affinity[8]. In ad-

Figure 5. Structure of 5'-phospho-uridine-2',3'-cyclic phosphate

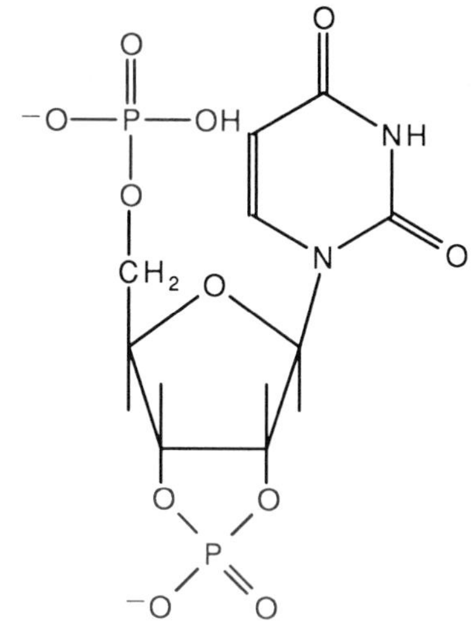

dition, a 5-fold increase in $k_{cat.}/K_m$ has been found for the hydrolysis of 5'-phospho-uridine 2',3'-cyclic phosphate (Figure 5) as compared to 2',3'-cyclic UMP, clearly supporting the existence of p_0[9]. Based on X-ray diffraction and model building studies it was proposed that p_0 is essentially formed by Lys-66[10].

THE p_2 BINDING SUBSITE

The first evidence for the existence of an additional phosphate-binding subsite (p_2) at the 3'-side of R_2 (see Figure 3) was the specific reaction of 6-chloropurine riboside 5'-monophosphate (cl6RMP) with ribonuclease A[11] (Figure 6). The reaction yielded a single major derivative (Derivative II) at the α-NH$_2$ of Lys-1, while reaction with the corresponding halogenated base (6-chloropurine) and nucleoside (6-chloropurine riboside) yielded several mono- and multi-derivatives. Moreover, reaction with cl6RMP was partially prevented by the presence of 5'-AMP in the reaction mixture. These

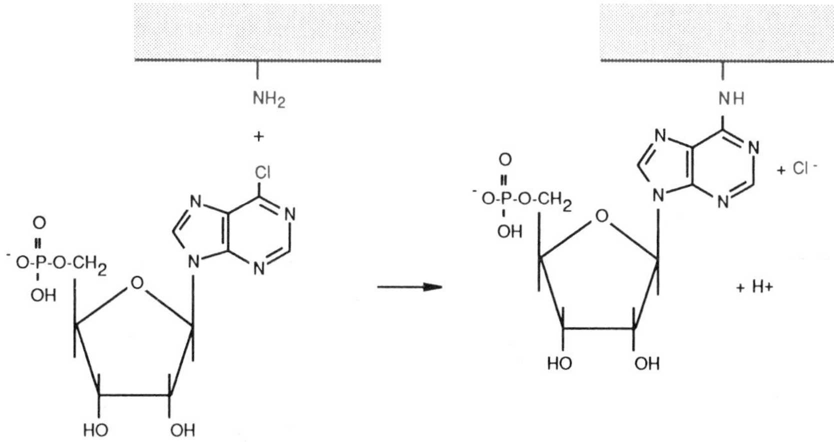

Figure 6. Reaction between 6-chloropurine riboside 5'-monophosphate and ribonuclease A at the α-NH$_2$ of Lys-1

results indicated that the reaction was an affinity labelling and that the phosphate was essential for the specificity. Derivative II exhibited practically unchanged kinetic constants with 2′,3′-cyclic CMP but the K_m for RNA was twice the native value.

The data were interpreted in terms of the existence of a phosphate binding subsite (p_2) in the N-terminal region of ribonuclease. Before the covalent labelling takes place, the phosphate of the cl⁶RMP binds in p_2 and conveniently orientates the reagent to allow the nucleophilic attack of the α-NH_2 on C-6. The binding of the substrate 2′,3′-cyclic CMP, just in $p_1R_1B_1$, is not hindered by the p_2 blocking group in Derivative II. However, binding of RNA, which has to occupy p_2 for an effective catalysis, is partially prevented in Derivative II.

The cl⁶RMP–ribonuclease A reaction has been studied using different purine and pyrimidine nucleotides as protecting agents[12]. The efficiency of preventing the reaction followed the order 3′-AMP > 5′-AMP = 5′-CMP > 3′-CMP, that would correspond to the extent of p_2 occupancy. Therefore 3′-AMP is specific for p_2, and will occupy $p_2R_2B_2$ (Figure 3), in accordance with the X-ray data[1], while 5′-AMP mainly binds at $B_2R_2p_1$, although at high concentrations it would also bind at p_2, probably at $B_3R_3p_2$. These conclusions were corroborated by ¹H-n.m.r. studies of Derivative II and complexes of ribonuclease A and nucleotides of different structures[7,13].

Irie and coworkers also provided evidence on the existence of p_2. They found that the inhibitory effect of pAp towards the hydrolysis of 2′,3′-cyclic CMP by ribonuclease A was about 10 times greater than that of 5′-AMP. This was explained by the additional binding of the 3′-phosphate of pAp at the p_2 binding site. Steady-state kinetic studies demonstrated that the $k_{cat.}$ values for the trinucleoside diphosphates UpApA and UpApG were 3–5-fold higher than those for the dinucleoside monophosphate UpA, suggesting that the additional pA or pG (occupying p_2B_3) enhanced the catalysis[14].

In order to identify the amino acid residues involved in the p_2 phosphate binding site and to differentiate it from p_1, chemical modification with pyridoxal phosphate was performed using 3′-AMP and 5′-AMP as protecting agents for p_2 and p_1, respectively[15]. Pyridoxal 5′-phosphate reacts with ribonuclease A to produce phosphopyridoxyl monoderivatives at Lys-1, Lys-7 and Lys-41 (Figure 7). As previously mentioned,

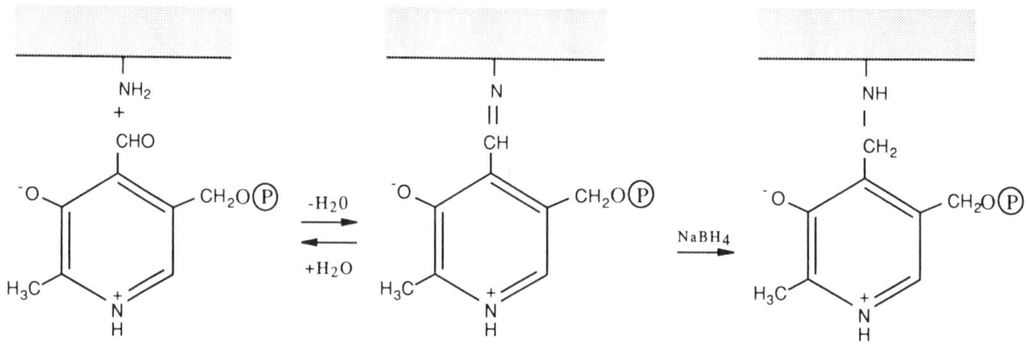

Figure 7. Reaction between pyridoxal phosphate and ribonuclease A
The reaction takes place at the NH_2 of certain lysines.

Lys-41 is close to the p_1 binding site, while computer graphics studies suggested that Lys-7 is close to p_2[4]. When either of the two nucleotides (3'-AMP or 5'-AMP) were present in the labelling reaction, the modification of both Lys-7 and Lys-41 was strongly prevented. These results and the kinetic studies of the derivatives supported the involvement of Lys-7 in p_2, and suggested that both phosphate-binding sites p_1 and p_2 are needed for the reaction in either lysine. Thus, the phosphate group of the label must interact in p_2 (near Lys-7) to direct the covalent reaction in Lys-41 (p_1), and the phosphate group of the label must interact in Lys-41 when the reaction is at Lys-7[15]. Additional evidence for the involvement of Lys-7 in p_2 was provided by kinetic studies with uridine oligonucleotides of different size and ribonuclease S', in which the Lys-7 was substituted by norleucine[16].

On the basis of computer graphics models Arg-10 is also a good candidate as a component of p_2. To study this possibility, chemical modification of Arg-10 was performed with cyclohexane-1,2-dione[15]. Modification was only decreased by 19% in presence of 5'-AMP whereas 3'-AMP, which specifically binds in p_2, fully protected Arg-10, thus supporting the involvement of this residue in p_2.

ADDITIONAL BINDING SUBSITES

In the early studies on nucleic acid-ribonuclease A binding it became apparent that multiple interactions existed. Research in this field was initiated in order to understand the helix-destabilizing activity of ribonuclease on native DNA. Jensen and von Hippel[17] showed that the protection, or site-covered size, was 10–12 nucleotides long when the protein binds to single-stranded DNA. Chemical modification experiments further demonstrated that these sites were primarily formed by lysines and arginines, and that they were not randomly distributed on the surface of the enzyme but had specific locations[18].

Important support for the existence of multiple subsites in ribonuclease A was provided by the crystallographic analysis of complexes between the protein and tetra-deoxyadenylic acid, $d(pA)_4$[19]. It was found that four $d(pA)_4$ molecules were partially bound to a single protein molecule. Ignoring the nucleotide units involved in inter-molecular interactions, it was seen that a total of 12 nucleotide units were interacting with the enzyme. These nucleotides trace out a near-continous path running through the active site, over the surface of the protein, and finally into the electropositive cluster on the "back" of the protein, thus forming a "virtual DNA strand". The binding between protein and nucleic acid is mainly an extended cation–anion interaction. Salt bridges are formed between phosphate groups and nine charged side chains (Figure 8). The only important interactions involving the bases occur at the active site. The lysine and arginine groups are in fact presented in a linear array over the surface of the molecule so that they are spatially complementary to the arrangement of phosphate groups along the course of a polynucleotide chain. The ribonuclease A structure would guide the single-stranded nucleic acid molecule through the active site cleft in an energy-efficient manner that does not perturb, but is in fact consistent with, the natural conformational preferences of DNA[20]. These authors stated that the same conclusions apply for the RNA binding since both ribonucleotides and deoxyribonucleotides interact with similar strength with ribonuclease, and single-stranded DNA is a good inhibitor of the ribonuclease activity. This structural model confirms the location of the

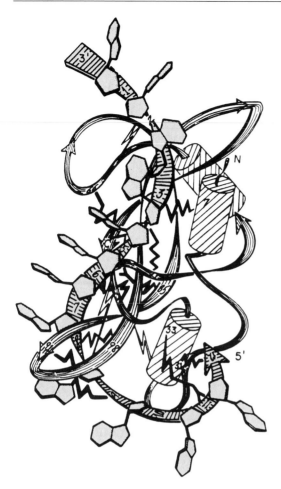

Figure 8. Schematic drawing of the model for the complex between ribonuclease A and d(pA)$_4$
The pattern of eight to nine electrostatic interactions that bind the polynucleotide to the surface of the protein is shown. These include salt bridges between phosphate groups and Lys-7, Lys-41, Lys-66, Arg-85, Arg-39, Lys-91, Lys-98, Arg-33 and Lys-31[20].

most studied subsites (p_2, p_1, and p_0) and predicts one additional nucleotide-binding site at the 3' end and eight at the 5' end.

Chemical modification with 6-chloropurine riboside 5'-monophosphate as well as model building and computer graphics studies using a complex of ribonuclease A and the pentanucleotide model pApUpApApG also served further to delineate the nucleotide binding subsites[4]. Essentially, five subsites were characterized, the three well-known sites at the active centre cleft (p_2, p_1, and p_0 with the corresponding sites for bases and riboses) plus two additional subsites: p_3 at the 3' end and $p_{1'}$ at the 5' end (Figure 9). Consistent with previous results, Lys-7 and Arg-10 are at p_2, Lys-41 at p_1 and Lys 66 at p_0. Lys-37 is involved in p_3 while Lys-104 is responsible for the phosphate binding at $p_{1'}$. These five positive regions nicely match with the five zones of anion interaction described by Matthew and Richards[21]. However, although there is agreement in the location of p_2, p_1 and p_0, McPherson and coworkers[20] assigned Arg-39 to $p_{1'}$ and no positive residue was located near p_3. These discrepancies may arise from the use of different methodologies but also suggest that alternative paths of binding may exist. Thus, binding at p_2, p_1 and p_0 sites is essential for catalysis with polynucleotide substrates but additional subsites could be differently occupied depending on the nucleic acid structure.

This conclusion is supported by the sequence analysis of the pancreatic ribonucleases of 40 mammalian species[22]. Lys-7, Arg-10, Lys-41 and Lys-66 are invariant residues. Lys-104 is highly conserved and in the few exceptions is replaced by arginine. Other lysines and arginines involved in the additional subsites are much more variable, indicating a less essential role of these subsites.

ROLE OF THE BINDING SUBSITES IN RIBONUCLEASE A

Some important properties of ribonuclease can be explained in terms of the multiplicity of binding subsites in the protein molecule.

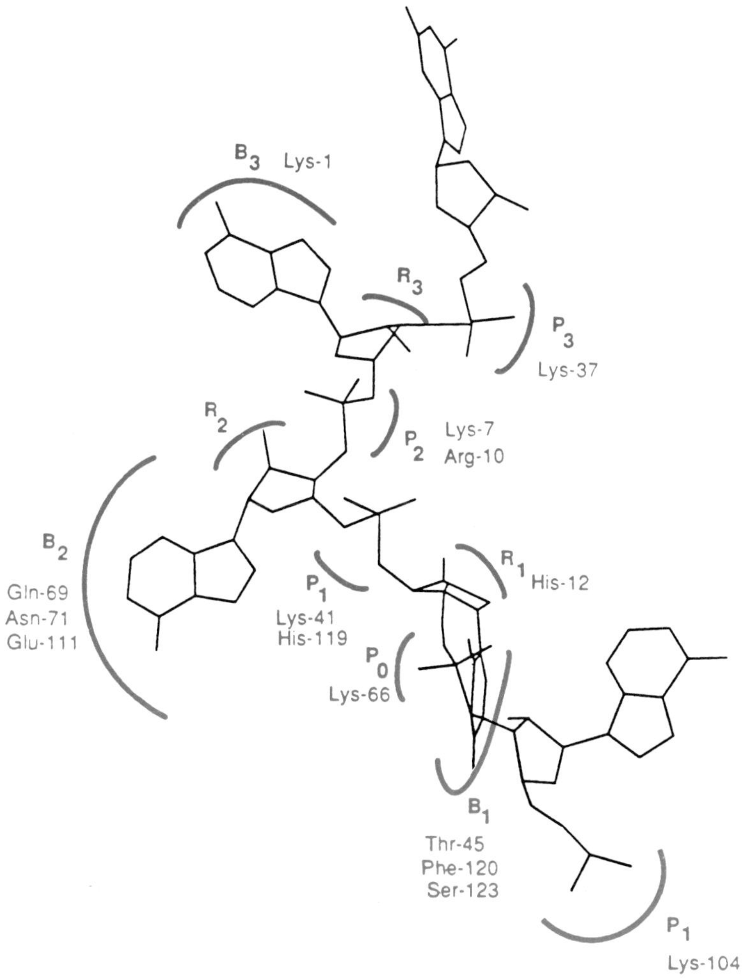

Figure 9. Schematic representation of the binding of an RNA fragment to ribonuclease A
See the legend to Figure 3 for the meaning of symbols. The spatial distribution of the RNA fragment corresponds to that of the pentanucleotide (pApUpApApG) model used in the computer graphics studies[4].

Specificity

The active centre subsites are responsible for the substrate specificity of ribonuclease A[1]. B_1 confers a practically absolute specificity for pyrimidines. R_1p_1 establishes the catalytic specificity. Although the enzyme binds ribonucleotides and deoxyribonucleotides equally well, only the former will be cleaved because a 2'-OH is necessary for the catalytic mechanism. The B_2 subsite exhibits preference for purines which results in higher $V_{max.}$ values for substrates with a purine in the 5' side of the phosphodiester bond. This activation occurs probably through a conformational change induced by the binding at B_2[7].

Catalytic efficiency

Activity towards oligonucleotides increases with the chain length of the substrate. Experiments with oligouridylic acids of increasing chain length up to five nucleotides indicate that the kinetic parameters are almost invariable for substrates with three or more nucleotides, suggesting that the number of subsites important for catalysis correspond to three nucleotides. However, $V_{max.}$ for polyuridylic acid is 3 to 20 times higher (depending on the assay conditions) than the $V_{max.}$ value for the oligonucleotide of three or four uridylic acid units[23]. Consequently, additional binding subsites have a significant role in the catalysis of long polynucleotides. It can be hypothesized that the occupation of the three nucleotide-binding subsites of the active centre cleft will enhance catalysis by favouring a correct alignment of the substrate and probably by inducing a conformational change on the protein, as discussed for the binding at B_2. Additional nucleotide-binding subsites would help catalysis of transphosphorylation of polynucleotides, probably by avoiding non-productive binding, which would be frequent with short-chain substrates. Data available suggest that a number of additional binding sites should be occupied by the polynucleotide to yield a significant increase in $V_{max.}$. This number should be clearly higher than that corresponding to five nucleotides[23].

To determine the minimum size of the polynucleotide chain for optimal catalysis, polycytidylic acid was incubated with a low concentration of ribonuclease A and the oligonucleotides produced by the reaction at different times were analyzed by h.p.l.c.[24]. During the first few minutes, the reaction only yields polynucleotide fragments longer than 10–20 nucleotides. Interestingly, after 10 minutes reaction there is a significant accumulation of oligocytidylic acid of six or seven residues. It appears that the enzyme prefers binding and cleavage of long substrates and that at least one of the ends of the phosphodiester bond should have six or seven nucleotides to be preferentially cleaved by ribonuclease.

These results and many others from the extensive ribonuclease A literature[1–3] demonstrate that after the transphosphorylation step, both fragments of the cleaved polynucleotide (one of them containing a 2',3'-cyclic phosphate terminus) leave the enzyme and are replaced by a new, long chain, substrate molecule. Subsequently, the shorter fragments will also be cleaved, and eventually the hydrolytic step occurs when most of the RNA has been cleaved in the transphosphorylation step. The multi-subsite structure of ribonuclease A gives an explanation of this phenomenon. The strong binding between a long RNA chain and a ribonuclease molecule takes place because

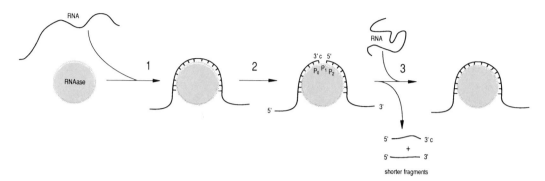

Figure 10. Model of the cleavage of an RNA chain by ribonuclease A that explains the preference of the enzyme for long polynucleotide substrates
The model is based on the co-operative binding between the multiple protein subsites and the phosphates of the polynucleotide. (1) A long RNA chain binds to ribonuclease. (2) Cleavage occurs in the active site resulting in the formation of two shorter oligonucleotide fragments, one of them ending with a 2′,3′-cyclic phosphate, indicated as 3′c. (3) The co-operative binding is weakened in the oligonucleotide fragments and this favours their replacement by a longer chain that fully occupies the ribonuclease A binding sites. After the transphosphorylation step (Figure 1), both fragments of the cleaved polynucleotide leave the enzyme and are replaced by a new, long-chain, substrate molecule. In subsequent reactions the shorter fragments will also be cleaved, and eventually the hydrolytic step occurs when most of the RNA has already been cleaved in the transphosphorylation step. Reproduced, with permission, from [20].

of the co-operativity of the multiple weak forces established between the polynucleotide and the subsites of the protein (Figure 10). When the chain is broken in the transphosphorylation step of the catalysis, the co-operativity is partially lost in the remaining shorter fragments. A new intact long chain, occupying most of the subsites, will then displace the fragments. Polynucleotides longer than the minimum size for optimal catalysis (probably longer than six or seven nucleotides) will be cleaved first and, subsequently, shorter fragments will be progressively cleaved in the order of maximum to minimum occupancy of the protein subsites. The hydrolytic step occurs when most of the suitable phosphodiester bonds are already broken because $k_{cat.}/K_m$ for this step is several orders of magnitude lower than that for the transphosphorylation step (Table 1).

Helix destabilizing activity

Ribonuclease A destabilizes the native DNA conformation, i.e. it lowers the temperature (T_m) of the double-helix-to-coil melting transition. This property results from its preferential binding to the single-stranded polynucleotide chain. It is believed that the protein "traps" single-stranded sequences transiently exposed by thermal fluctuations. Jensen and von Hippel[17] demonstrated that charge–charge interactions contribute markedly to the free energy of the binding, which doubtless involves the binding of negative charged DNA phosphates with positive groups of the enzyme. Karpel and coworkers[18] have convincingly demonstrated that removal of the positive group of lysines by chemical modification destroys the ability of ribonuclease A to lower the T_m of poly d(A-T). Modification of arginines also decreases the T_m-depressing activity of the enzyme. Ribonuclease S (ribonuclease A cleaved between residues 20 and 21) exhibits similar effects on T_m to those of the native enzyme, but S-protein (the fragment of ribonuclease S comprising residues 21–124), which only lacks three charged residues (Lys-1, Lys-7 and Arg-10) has no effect on T_m. This suggests that the specific location of the positive groups (namely the substrate-binding subsites) on the ribonuclease surface is largely responsible for the protein's DNA-melting ability.

It has been argued that the T_m-depressing activity of ribonuclease A may have little physiological relevance. However, although ribonuclease is unable to unstack bases of native DNA, it readily destroys the hairpin structures of single-stranded DNA[17]. We postulate that this ability may have a physiological role in the intestinal hydrolysis of RNA. Ribonuclease will first destroy the secondary structure of RNA, and then will carry out a complete digestion of the polynucleotide chain.

THE MULTIPLE SUBSITE STRUCTURE OF OTHER NUCLEASES AND NUCLEIC ACID-BINDING PROTEINS

Subsite multiplicity has also been defined for several microbial ribonucleases. The most studied is ribonuclease T1 (EC 3.1.27.3), extracted from *Aspergillus oryzae*. The enzyme performs a two-step endonucleolytic cleavage of RNA to 3'-phosphomono- and oligonucleotides ending in Gp, with 2',3'-cyclic phosphate intermediates. It consists of a chain of 104 amino acid residues, with two disulphide bridges, and with many negative charges that result in a pI of 2.9^{25}.

Ribonuclease T1 subsites were defined as enzyme interactions with the nucleoside and phosphate moieties for the polymeric sequence $..._2N_{-2}p_{-1}N_{-1}p\text{-}G\text{-}p_1\text{-}N_1\text{-}p_2\text{-}N_2...$ [26]. The guanine moiety (G) interacts at the primary recognition site. Additional subsites were found for p_1, N_1 and $_1N$. Studies of the transphosphorylation activity of ribonuclease T1 with oligoinosinic acids of different chain length indicated an increase of $V_{max.}$ as the number of residues increases up to three or four units[27]. The $V_{max.}$ value did not increase further with polyinosinic acid as substrate, indicating that additional binding subsites for long polynucleotide chains are probably absent from this protein. Similar transphosphorylation results were found for the related microbial ribonucleases ribonuclease St from *Streptomyces erythreus* and ribonuclease Ms from *Aspergillus saitoi*. We believe that the absence of additional binding subsites may be

Table 1. $k_{cat.}/K_m$ values for the transphosphorylation[a] and hydrolysis[b] reactions catalysed by ribonuclease A[1,11,14]

Substrate	pH	$k_{cat.}/K_m$
CpA	7.0	3000[a]
CpC	7.0	60[a]
ApC>p	7.0	12[b]
2',3'-cyclic CMP	7.0	1.6[b]
UpApG	5.0	97500[a]
CpA	5.0	12400[a]
2',3'-cyclic CMP	5.0	180[b]

related, at least in ribonuclease T1, to the lack of positive centres on the protein surface since one arginine and one lysine are the only amino acid residues with positive charge in its primary structure.

Another component of the family of microbial ribonucleases is barnase, the extracellular ribonuclease produced from *Bacillus amyloliquefaciens*, which shows substrate specificity and tertiary structure related to ribonuclease T1[28]. Barnase is a small protein of 110 amino acid residues with no disulphide bridges. In contrast to ribonuclease T1, barnase is very rich in residues with positively charged side chains (eight lysines and six arginines). Also in contrast to ribonuclease T1, the activity with polynucleotides is much higher than with shorter substrates. It can be hypothesized, in analogy to ribonuclease A, that the positive charges of barnase form additional centres of recognition for the phosphates of the substrate that would minimize the non-productive binding for the long polynucleotides and, therefore, would increase the catalytic efficiency towards these substrates.

Deoxyribonuclease I is another well-studied nuclease, which degrades double-stranded DNA to yield 5' oligonucleotides, with little base or sequence specificity. It constitutes another example of a protein that binds nucleic acid through multiple interactions mainly of electrostatic origin[29].

In recent years, several proteins that bind single-stranded DNA, of prokaryotic and eukaryotic origin, have been characterized (see [30] for a review). They constitute a structurally heterogeneous group, although some general features are the existence both of basic amino acids to neutralize phosphates and of aromatic residues that interact with the purines and pyrimidines. In general these proteins are larger and more complex than the ribonucleases. Most of them are probably involved in protein–protein interactions and fulfil important roles in DNA processing such as in the replicative process. The large amount of information available on the binding properties of ribonuclease A and the variety of methodologies used in its study will undoubtedly be of application in the research of this more complex group of proteins.

SUMMARY AND CONCLUSIONS

- Ribonuclease A binds nucleic acids through multiple electrostatic interactions between the phosphates of the polynucleotide and the positive groups (side chains of lysines and arginines) of the protein subsites. The bases only play a significant role in the binding at the active site. The active centre $p_1R_1B_1$ sites determine the specificity of the catalytic cleavage.

- The phosphate-binding subsites p_2 (Lys-7 and Arg-10), p_1 (Lys-41, His-12 and His-119) and p_0 (Lys-66) are essential for an effective catalysis and are conserved in all mammalian pancreatic ribonucleases. Additional phosphate-binding subsites confer further catalytic efficiency, probably by avoiding non-productive binding. The minimum chain size for optimum catalysis is probably longer than six or seven nucleotides. The full occupancy of binding sites by the long chain polynucleotides would explain the preference of the enzyme for these substrates.

- The multiplicity of binding subsites is responsible for the helix-destabilizing activity of ribonuclease A. Its capacity for destroying the secondary structure of single-stranded nucleic acids may be of importance for the complete hydrolysis of RNA in the digestive tract.

● A large variety of proteins, with very different structures and functions, interact with nucleic acids. An analysis of their binding properties shows that there is no general model for protein–nucleic acid interaction. However, the vast amount of work on the ribonuclease A binding subsites should serve as a model for the study of the binding properties of many other proteins that recognize nucleic acids.

The experimental part of this paper performed by our research group has been supported by grants (1716/82, PB85-97 and PB88-232) from the Ministerio de Educación y Ciencia. We also thank Fundació M.F. de Roviralta for grants for the purchase of equipment.

REFERENCES

1. Richards, F.M. & Wyckoff, H.W. (1971) Bovine pancreatic ribonuclease. *The Enzymes* **4**, 647–806

2. Blackburn, P. & Moore, S. (1982) Pancreatic ribonuclease. *The Enzymes* **15**, 317–433

3. Eftink, M.E. & Biltonen, R.L. (1987) Pancreatic ribonuclease A: the most studied ribonuclease, in *Hydrolytic Enzymes*, chapter 7, pp. 333–376, Elsevier Science Publishers, Amsterdam, New York and Oxford

4. de Llorens, R., Arús, C., Parés, X. & Cuchillo, C.M. (1989) Chemical and computer graphics studies on the topography of the ribonuclease A active site cleft. A model of the enzyme-pentanucleotide substrate complex. *Protein Eng.* **2**, 417–429

5. Wlodawer, A. (1985) in *Biological Macromolecules and Assemblies*, volume 2, chapter 9, pp. 393–439, John Wiley and Sons, New York

6. Usher, D.A., Richardson, D.I. & Eckstein, F. (1970) Absolute stereochemistry of the second step of ribonuclease action. *Nature (London)* **228**, 663–665

7. Arus, C., Paolillo, L., Llorens, R., Napolitano, R. & Cuchillo, C.M. (1982) Evidence on the existence of a purine ligand induced conformational change in the active site of bovine pancreatic ribonuclease A studied by proton nuclear magnetic resonance spectroscopy. *Biochemistry* **21**, 4290–4297

8. Sawada, F. & Irie, M. (1969) Interaction of uridine 2′(3′),5′-diphosphate with ribonuclease A and carboxymethylribonuclease A. *J. Biochem. (Tokyo)* **66**, 415–418

9. Li, J. R.-T. & Walz, G.F. (1974) A steady-state kinetic study of the ribonuclease A catalyzed hydrolysis of uridine-2′:3′(cyclic)-5′-diphosphate. *Arch. Biochem. Biophys.* **161**, 227–233

10. Mitsui, Y., Urata, Y., Torii, K. & Irie, M. (1978) Studies on the binding of adenylyl-3′,5′-cytidine to ribonuclease. *Biochim. Biophys. Acta* **535**, 299–308

11. Parés, X., Llorens, R., Arús, C. & Cuchillo, C.M. (1980) The reaction of bovine pancreatic ribonuclease A with 6-chloropurineriboside 5′-monophosphate: evidence on the existence of a phosphate-binding sub-site. *Eur. J. Biochem.* **105**, 571–579

12. Richardson, R.M., Parés, X. & Cuchillo, C.M. (1990) Chemical modification by pyridoxal 5′-phosphate and cyclohexane-1,2-dione indicates that Lys-7 and Arg-10 are involved in the p_2 phosphate-binding subsite of bovine pancreatic ribonuclease A. *Biochem. J.* **267**, 593–599

13. Alonso, J., Paolillo, L., D'Auria, G., Nogués, M.V. & Cuchillo, C.M. (1989) [1]H-n.m.r. studies on the existence of substrate binding sites in bovine pancreatic ribonuclease A. *Int. J. Peptide Protein Res.* **34**, 66–69

14. Irie, M., Watanabe, H., Ohgi, K., Tobe, M., Matsumura, G., Arata, Y., Hirose, T. & Inayama, S. (1984) Some evidence suggesting the existence of P$_2$ and B$_3$ sites in the active site of bovine pancreatic ribonuclease A. *J. Biochem. (Tokyo)* **95**, 751–759

15. Richardson, R.M., Parés, X., de Llorens, R., Nogués, M.V. & Cuchillo, C.M. (1988) Nucleotide binding and affinity labelling support the existence of the phosphate-binding subsite P$_2$ in bovine pancreatic ribonuclease A. *Biochim. Biophys. Acta* **953**, 70–78

16. Irie, M., Ohgi, K., Yoshinaga, M., Yanagida, T., Okada, Y. & Teno, N. (1986) Roles of lysine 1 and lysine 7 residues of bovine pancreatic ribonuclease in the enzymatic activity. *J. Biochem. (Tokyo)* **100**, 1057–1063

17. Jensen, D.E. & von Hippel, P.H. (1976) DNA "melting" proteins: I. effects of bovine pancreatic ribonuclease binding on the conformation and stability of DNA. *J. Biol. Chem.* **251**, 7198–7214

18. Karpel, R.L., Merkler, D.J., Flowers, B.K. & Delahunty, M.D. (1981) Involvement of basic amino acids in the activity of a nucleic acid helix-destabilizing protein. *Biochim. Biophys. Acta* **654**, 42–51

19. McPherson, A., Brayer, G. & Morrison, R. (1986) Structure of the crystalline complex between ribonuclease A and d(pA)$_4$. *Biophys. J.* **49**, 209–219

20. McPherson, A., Brayer, G., Cascio, D. & Williams, R. (1986) The mechanism of binding of a polynucleotide chain to pancreatic ribonuclease. *Science* **232**, 765–768

21. Matthew, J.B. & Richards, F.M. (1982) Anion binding and pH-dependent electrostatic effects in ribonuclease. *Biochemistry* **21**, 4989–4999

22. Beintema, J.J., Schüller, C., Irie, M. & Carsana, A. (1988) Molecular evolution of the ribonuclease superfamily. *Progr. Biophys. Mol. Biol.* **51**, 166–192

23. Irie, M., Mikami, F., Monma, K., Ohgi, K., Watanabe, H., Yamaguchi, R. & Nagase, H. (1984) Kinetic studies on the cleavage of oligouridylic acids and poly U by bovine pancreatic ribonuclease. *J. Biochem. (Tokyo)* **96**, 89–96

24. Guasch, A., Nogués, M.V. & Cuchillo, C.M., unpublished work

25. Takahashi, K. & Moore, S. (1982) Ribonuclease T1. *The Enzymes* **15**, 435–469

26. Osterman, H.L. & Walz, F.G. (1979) Subsite interactions and ribonuclease T1 catalysis: kinetic studies with ApGpC and ApGpU. *Biochemistry* **18**, 1984–1988

27. Watanabe, H., Ando, E., Ohgi, K. & Irie, M. (1985) The subsite structures of guanine-specific ribonucleases and a guanine-preferential ribonuclease: cleavage of oligoinosinic acids and poly(I). *J. Biochem. (Tokyo)* **98**, 1239–1245

28. Mossakowska, D.E., Nyberg K. & Fersht, A.R. (1989) Kinetic characterization of the recombinant ribonuclease from *Bacillus amyloliquefaciens* (barnase) and investigation of key residues in catalysis by site-directed mutagenesis. *Biochemistry* **28**, 3843–3850

29. Suck, D. & Oefner, C. (1986) Structure of DNAase I at 2.0 Å resolution suggests a mechanism for binding to and cutting DNA. *Nature (London)* **321**, 620–625

30. Chase, J.W. & Williams, K.R. (1986) Single-stranded DNA binding proteins required for DNA replication. *Annu. Rev. Biochem.* **55**, 103–136

9

Metabolic studies using ¹³C nuclear magnetic resonance spectroscopy

Ronnitte Badar-Goffer and Herman Bachelard

NMR Biospectroscopy Group, Department of Physics, University of Nottingham, University Park, Nottingham NG7 2RD, U.K.

INTRODUCTION

In the first half of this century, materials labelled with radioactive or non-radioactive tracers were slowly introduced into biological studies. One of the first biological applications of the use of a radioactive tracer was in plants, where uptake of lead was studied[1]. Radioactive isotopes that are more relevant to general metabolic studies became available later and use of the short-lived ¹¹C isotope was first demonstrated by Kamen's group in 1940. This had the disadvantage at that time of its very short half-life of 20 minutes (although it is now appreciated for its value in positron emission tomography) and it was rapidly replaced as a general tracer by the long-lived ¹⁴C isotope. It was only after the second world war that there was a general increase in the use of isotopic methods; ¹⁴C tracer studies were extensively applied to the elucidation and characterization of metabolic pathways and measurement of flux rates (see [2]).

One stable isotope of carbon (¹³C) has a nuclear spin of 1/2 which allows for nuclear magnetic resonance detection. Nuclear magnetic resonance (n.m.r.) spectroscopy has several appealing features for applications to metabolic studies, the most obvious of which is the non-invasiveness of the technique, which allows for compounds to be detected in their natural environment. The nuclei that are most commonly used in n.m.r. for metabolic studies are: ¹H, ³¹P and ¹³C. ¹H and ³¹P are naturally-abundant isotopes, and therefore the most common methods of study involve examining differences in the natural abundance spectra under various metabolic states. In contrast, ^{13}C has a natural abundance of only 1.1%, and is an insensitive nucleus (in n.m.r. terms, relative to ^{1}H). Such disadvantages normally make it difficult and limited for studies on endogenous metabolites unless they occur in large amounts. However, the low natural abundance can be an advantage in that ^{13}C-enriched precursors can be used for metabolic pathway mapping with little or no background interference from endogenous metabolites. Thus studies analogous to those earlier carried out with ^{14}C tracers can now be performed by ^{13}C-n.m.r.

A further advantage is that ^{13}C has a very large chemical shift range (200 p.p.m., as compared with 10 p.p.m. for ^{1}H and 30 p.p.m. for ^{31}P), thus allowing for individual carbon atoms to be detected within the same molecule. (Chemical shift range refers to the frequency "window" of the resonances of the nucleus: a large range gives a wide window providing good resolution, whereas a small range compresses the resonances, thus limiting resolution; see Figure 1.) The entry of a ^{13}C-labelled precursor into different metabolites can be followed simultaneously without the need for chemical isolation or purification; the individual carbons which are specifically labelled in each individual metabolite can be resolved, providing sufficient signal/noise ratio is

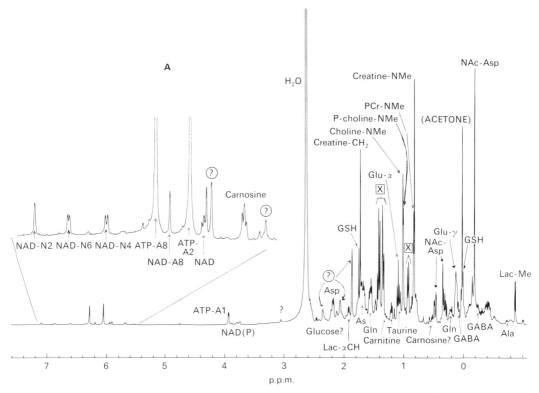

Figure 1 (continued opposite). Comparison of the chemical shift ranges of ^{1}H (10 p.p.m., A), ^{31}P (30 p.p.m., B) and ^{13}C (200 p.p.m., C) n.m.r. spectra of guinea pig brain extracts
In A, part of the ^{1}H spectrum has been expanded to show the resolution that can be obtained. This can also be seen in the ^{13}C spectrum of Figure 5(2) below, where the amino acids resonances are well resolved. Data are from our unpublished work (A, J. Feeney & H.S. Bachelard; B, P.G. Morris & H.S. Bachelard; C, R.S. Badar-Goffer).

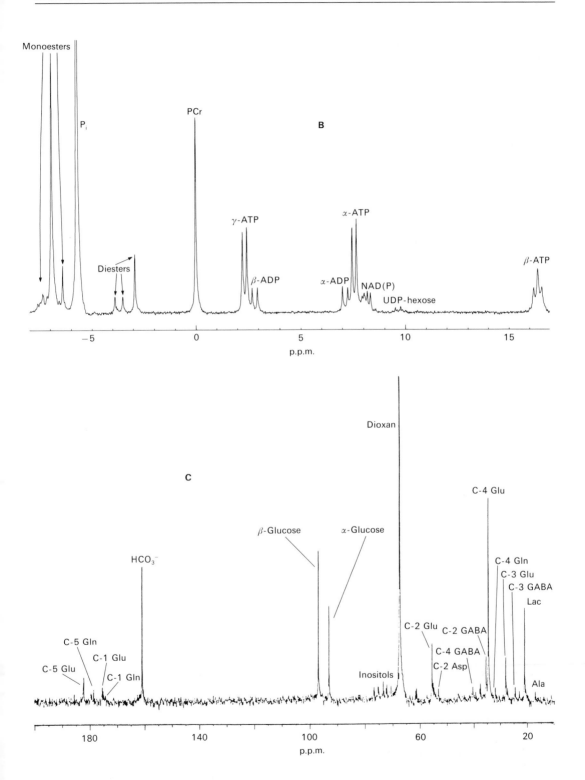

Figure 1 (continued)

obtained. Classical methods using ^{14}C-labelled precursors remain much more sensitive than ^{13}C-n.m.r., due to the relative insensitivity of the latter technique. However, it is the exceptional chemical resolution that renders it uniquely suitable for metabolic studies.

Although ^{13}C is only 1.1% naturally abundant, compounds that occur at high levels such as mobile lipids and glycogen can be detected non-invasively *in vivo*. Therefore, changes in the natural abundance spectra following metabolic perturbation, or comparative studies on different tissues, can yield useful information.

For compounds occurring at lower levels, the precursors used for labelling are usually over 90% enriched; therefore the information obtained differs from that available from ^{14}C, which is normally used in tracer amounts. Using ^{13}C-labelled precursors containing enriched specified carbon atoms, it is possible to follow the incorporation of label into products either *in vivo* or *in vitro*. The information that can be obtained will ultimately depend on the signal-to-noise ratio and the spectral resolution. In certain instances, particularly *in vivo*, signals will be broad allowing only limited information to be obtained. Thus *in vivo* studies are usually accompanied by *in vitro* studies performed on tissue extracts in which significantly superior resolution is obtained, and therefore far more information can be elicited. These include metabolites not visible in the *in vivo* spectra, calculation of %^{13}C enrichment and ^{13}C–^{13}C splitting patterns, namely isotopomer analysis, a technique which is evolving rapidly. All of these will be discussed in detail below.

To overcome the sensitivity problems of ^{13}C-n.m.r. and still take advantage of the ability of n.m.r. to follow metabolism *in vivo* by using ^{13}C-labelled precursors, an indirect method, of ^{1}H observation of ^{13}C labelling, has been devised and is described below.

Clinical applications of ^{13}C-n.m.r., and human studies using ^{13}C-labelled precursors or natural abundance ^{13}C, are now taking place with hopes to further the understanding of inherited enzyme deficiencies and the development of diagnostic tests for these and other metabolic disorders. These studies are referred to at the end of the Review.

In this article we have selected examples to illustrate the principles and techniques, which are evolving; it is therefore not intended to be a comprehensive review of the complete literature published in this field of study.

NATURAL ABUNDANCE STUDIES

The main resonances observed in a straightforward *in vivo* ^{13}C spectrum are those that occur at sufficiently high levels to be detected without labelling from ^{13}C precursors. These include lipids (triacylglycerols and phospholipids) and carbohydrates (glucose and glycogen). In preliminary studies using radiofrequency surface coils on liver and adipose tissue *in vivo*, constituents of the spectra were identified by comparison with those from the excised organs or in tissue extracts. Changes in the natural abundance spectra were observed in the living rat following chronic modifications of the diet. The researchers found it possible to distinguish the polyunsaturated olefinic ^{13}C peak at 128.5 p.p.m. from the peak at 130 p.p.m. derived from other unsaturated fatty acids, thus exhibiting the excellent chemical resolution inherent in the technique[3]. Chronic restriction of essential (polyunsaturated) fats from the diet resulted in a decrease in the polyunsaturated fatty acid resonance at 128.5 p.p.m. in adipose tissue spectra from the living rat. Similarly, a chronic supplementation of the diet with essential fats resulted in an increase of the resonance at 128.5 p.p.m.[3].

Natural abundance ^{13}C-n.m.r. was used to monitor glycogen in glycogen storage disease in rat liver by using surface coils[4]. The concentration of glycogen determined by n.m.r. and by chemical analysis was similar, and approximately three times that found in normal rats. Starvation did not reduce the glycogen content of the livers with glycogen storage disease, whereas it reduced the signal below detectability in normal rats. Difference spectra (Figure 2), obtained by subtracting the spectra of starved normal rats from those of fed rats with glycogen storage disease, gave spectra similar in appearance to that of purified glycogen. In this study glycogen was found to be fully visible both *in vivo* and in tissue extracts by ^{13}C-n.m.r.

Studies of ^{13}C natural abundance in the brain were first performed by Barany *et al.*[5]. Following assignment of many of the multitude of resonances observed in ^{13}C-n.m.r. spectra of neutralized perchloric acid extracts, studies on the effects of halothane,

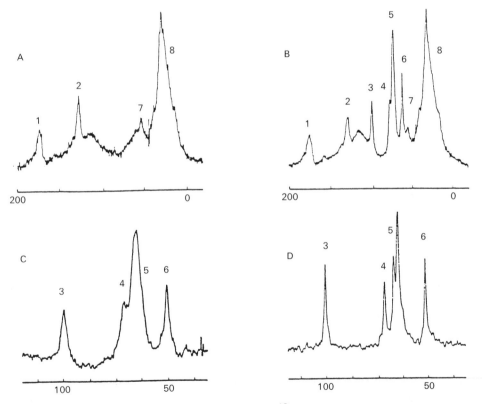

Figure 2. Proton-decoupled natural abundance ^{13}C-n.m.r. spectra of starved rat liver *in vivo*

A, normal animals; B, phosphorylase-deficient animals; C, difference spectrum, where A has been subtracted from B; D, pure glycogen. The resonances attributable to glycogen (3, C-1; 4, C-4; 5, C-2, C-3 and C-5; 6, C-6) are all present in B, but absent from A. The difference spectrum (C) consists almost solely of glycogen. Other resonances (1, 2, 7 and 8) were attributed to carbons of proteins and lipids. Data with permission from Stevens *et al.*[4].

a common anaesthetic, on the whole brain *ex vivo* were carried out. Rat brains were excised and incubated in screw-capped vials with a high concentration of halothane (100 mmol/g of brain) at 37 °C for 10 min, whereupon the brains were transferred to the n.m.r. tubes. The controls were excised brains incubated similarly in the absence of halothane. In the control brain a resonance at 30.5 p.p.m., which corresponds to the -$(CH_2)_n$- units of phospholipids, was barely observable whereas it significantly increased with halothane[5]. Although such studies on excised (very ischaemic) brain would be frowned upon by biochemists, they did serve to illustrate the possibility of following quantitative changes in phospholipids in the intact tissue by natural abundance ^{13}C-n.m.r. spectroscopy.

In summary, natural abundance ^{13}C-n.m.r. spectroscopy can be a valuable tool to investigate changes that occur in the more concentrated cellular components, and has already provided important information, particularly in the study of glycogen storage diseases. Examples of further developments of this aspect in human studies will be presented subsequently in this Review.

USE OF ^{13}C-ENRICHED PRECURSORS

Administration of ^{13}C-labelled precursors either *in vivo* or *in vitro* can provide an elegant way to follow the entry of label into a large variety of metabolic products. The background signal can be subtracted from the spectra accumulated after the addition of the labelled precursor, resulting in spectra that will contain signal only from the enriched species. We and others have used this technique in perfused heart, liver and superfused brain slices. Again the only limitation of the technique is the insensitivity of ^{13}C-n.m.r.; label can only be accurately observed in those metabolites that are both highly labelled and at relatively high levels in the tissue.

Following the administration of selected precursors, gluconeogenesis and the pentose phosphate pathway were studied simultaneously in the liver, using a combination of ^{14}C tracers and ^{13}C-n.m.r. spectroscopy[6]. Thus gluconeogenesis and activity through the pentose phosphate shunt were compared using [2-^{13}C]glycerol at normal substrate levels and [2-^{14}C]glycerol in tracer amounts. Labelling patterns from glycerol to glycerol 3-phosphate and into the C-2 and C-5 positions of glucose were followed. The activity of the pentose phosphate pathway was assessed from the ratio of the C-2 to C-5 labels on glucose, since operation of this pathway causes a loss of carbon label at C-2 relative to C-5. The relative concentrations of ^{13}C label at specific carbon atoms were measured from the n.m.r. spectra and were found to be in agreement with the ^{14}C isotopic distributions measured in the same extracts. Cohen[7] studied the effect of insulin on hepatic metabolism of [2-^{13}C]pyruvate, [1,2-^{13}C]ethanol and NH_4^+. ^{13}C-labelled glycogen was found to increase synchronously with the synthesis of ^{13}C-labelled glucose. She also demonstrated[7] that ^{13}C-labelled citrate can provide a measure for intracellular Mg^{2+}. (Chelation of the Mg^{2+} causes a change in the chemical shift of the citrate resonance: the extent of the change in chemical shift is related to the available concentration of Mg^{2+}.) In this study, the relative ^{13}C enrichments on various carbon atoms of specific metabolites were described, but the absolute enrichments (i.e. analogous to absolute specific activities) were not measured.

DETERMINATION OF ABSOLUTE % ^{13}C ENRICHMENT

We have been investigating the metabolism of glucose and acetate in guinea-pig brain slices using ^{13}C-n.m.r. spectroscopy with the objective of following the metabolic relationships between neurones and glial cells. The basis of this approach is derived from earlier ^{14}C tracer studies on metabolic compartmentation between neurones and glia[8]. In addition to following labelling patterns in selected metabolites, we were able to measure the absolute ^{13}C enrichments of individual carbon atoms in those meta-bolites. These studies were performed on neutralized perchloric acid tissue extracts where the absolute ^{13}C enrichments were calculated on the basis of quantification of the resonances in the spectra compared with an internal dioxan standard, and direct measurement of the total pool size of the metabolite from amino acid analyses and enzymic determinations[9].

The total amount of ^{13}C in a particular resonance is given by:

$$[^{13}C]_{m(total)} = \frac{[^{13}C]_D \, (area)_m \, (SF)}{(area)_D \, (SF)_D} \qquad \text{(equation 1)}$$

where (area) = measured area beneath the ^{13}C resonance,
(SF) = saturating factor of the resonating species, if unsaturated,
m = metabolite, and
D = dioxan standard.

From this, the absolute % ^{13}C enrichment is calculated as:

$$\% \text{ enrichment} = \frac{([^{13}C]_{m(total)} - [^{13}C]_{m(n.a.)})}{[m]} \qquad \text{(equation 2)}$$

where $[^{13}C]_{m(total)}$ = total amount of ^{13}C (equation 1),
$[^{13}C]_{m(n.a.)}$ = naturally abundant ^{13}C in the metabolite, and
[m] = pool size of the metabolite from chemical or enzymic analysis.

In this way it becomes possible to follow the flux of ^{13}C from each individual position on one molecular species to another, e.g. from C-4 on glutamate or glutamine to the directly related C-2 position on 4-aminobutyrate (GABA)[9].

Another method for measuring absolute enrichment of ^{13}C comes from the use of gas chromatography-mass spectrometry (g.c.-m.s.). In a study of the metabolic path-ways involved in glycogen repletion in the liver *in vivo*, Kalderon et al.[10] infused, intra-intestinally into rats, a mixture of [U-^{13}C]glucose and unlabelled glucose, in order to minimize the probability of two labelled C_3 intermediates combining together to form a glucose molecule. The livers were then removed and extracted, and g.c.-m.s. analyses were performed to determine the ^{13}C enrichment of the glucose samples derived from liver glycogen. G.c.-m.s. analyses were also performed on other con-stituents of the extracts, including gluconeogenic precursors, lactate and alanine, glu-tamate and aspartate. Using this approach the investigators[10] concluded that only 35% of the newly-synthesized liver glycogen originates from the direct conversion of glucose into glycogen and that the other two-thirds of the glycogen is derived from the indirect pathway, i.e. conversion of glucose to C_3 units, mainly lactate and alanine, and subsequently via gluconeogenesis and glucose 6-phosphate to glycogen. This group (see Lapidot, 1990)[11] has since used g.c.-m.s. extensively to complement all their ^{13}C metabolic studies, some of which will be described later.

It is important to note that the two techniques are complementary: g.c.-m.s. gives absolute and highly accurate quantification of ^{13}C and ^{12}C on the whole molecule, but does not distinguish between the different carbon atoms within that molecule. ^{13}C-n.m.r. gives the amount of ^{13}C on each individual carbon atom, and therefore the relative distribution of ^{13}C between those carbon atoms, but absolute quantification is difficult. Ideally, both techniques should be used in combination.

It is also possible to determine the fractional ^{13}C-enrichment of a compound by using proton n.m.r. spectroscopy on extracts prepared from the labelled tissues. The basis for this determination is the "splitting patterns" obtained if a proton is attached to a ^{12}C or a ^{13}C carbon. The proton resonances from those coupled to ^{13}C will be symmetrically displaced about the central resonance of the protons attached to ^{12}C with a characteristic J_{CH} (coupling constant). However, because of the complexity of the proton spectra and many overlapping resonances due to the narrow chemical shift range, it has so far proven possible only with relatively few intermediates, such as lactate, which are reasonably well resolved (Figure 3). This approach has also been used on isolated, purified constituents , e.g. to determine the ^{13}C fractional enrichment of infused plasma glucose (see below)[11,12].

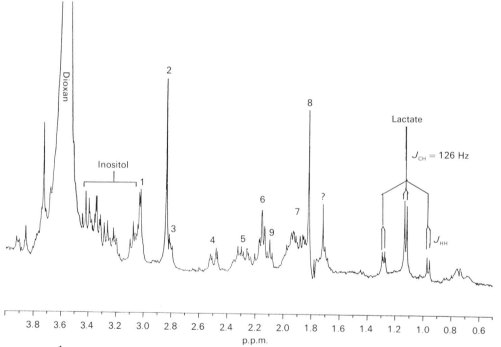

Figure 3. ^{1}H-n.m.r. spectrum of a brain extract after labelling with ^{13}C-precursors ([2-^{13}C]acetate plus [1-^{13}C]glucose), and depolarization with 40 mM-K$^+$

The lactate, which increases on depolarization, is well resolved; measurement of areas under the three resonances (main plus satellites) enables accurate quantification of the ^{13}C present in the lactate. Other resonances are: 1, -N(CH$_3$)$_3$ of choline-containing compounds; 2, -CH$_3$ groups of creatine and creatine phosphate; 3, γ-CH$_2$ of 4-aminobutyrate; 4, β-CH$_2$ of aspartate; 5, β-CH$_2$ of N-acetylaspartate; 6, γ-CH$_2$ of glutamate; 7, β-CH$_2$ of glutamate; 8, CH$_3$ of N-acetylaspartate; 9, γ-CH$_2$ of 4-aminobutyrate and (?) CH$_3$ of acetate (tentative). The data are from work in progress in our laboratories.

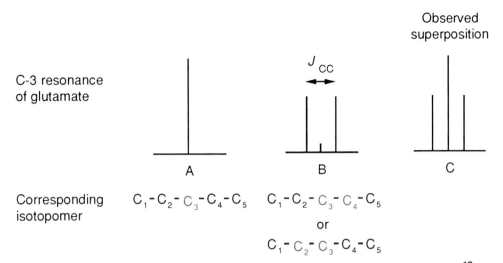

Figure 4. Diagrammatic representation of isotopomers of glutamate labelled with ^{13}C in the -CH$_2$- positions of a proton-decoupled ^{13}C-n.m.r. spectrum
The triplet (C) arises from superposition of the singlet C-3 resonance (A) and the doublet from the isotopomers with adjacent labelled carbons (B). In this case, C-3 + C-4 labelling cannot be distinguished from C-3 + C-2 labelling.

ISOTOPOMER ANALYSIS

The key to this type of analysis is the possibility of distinguishing between molecules of the same compound on which only one atom contains an isotope and those having atoms labelled with two or more of the same isotope (isomers + isotopes = "isotopomers"). We have seen that ^{13}C-n.m.r. spectroscopy has the advantage of a very wide chemical shift range which enables the resolution of each carbon atom in the same intermediate. If a molecule contains n carbon atoms the possible number of isotopomers is 2^n. In principle, analysis and quantification of the isotopomers in a ^{13}C-labelling study can provide much more information than can be derived from measurement of the % ^{13}C enrichments of each carbon atom[13]. In practice, it is complex if there are many isotopomers in the spectrum.

If a ^{13}C atom has an adjacent ^{13}C, its resonance will be split and displaced from the singlet arising from a single ^{13}C resonance, with a characteristic spin-spin coupling constant, J_{CC}. For example, the resonance of a mixture of glutamate labelled solely on C-4 and glutamate labelled on both C-4 and C-3 (i.e. both labels adjacent on the same molecule), will appear as a triplet. The ratio of the singlet (C-4$_{Glu}$) to the doublet (C-4,C-3$_{Glu}$) will depend directly on the actual concentrations of the two labelled species. The positions of the neighbouring adjacent labelled atoms can be determined from the J_{CC} constant, for example the J_{CC} (measured in Hz) of C-4,C-3 of glutamate will be different from that of C-4,C-5 of glutamate. If glutamate has three adjacent ^{13}C atoms, for example on C-3, C-4 and C-5, the resulting C-4 resonance will contain four peaks (a doublet of doublets). If the ^{13}C atoms are not directly adjacent to each another, for example C-4 and C-2 of glutamate, the spectrum will not reveal this because the long range coupling constants are too small to be detected. Further, the labelled species [2,3-^{13}C]glutamate and [3,4-^{13}C]glutamate cannot be resolved at C-3 and the two doublets will overlap (Figure 4).

An example of the application of isotopomer analysis is the use of specifically ^{13}C-labelled precursors for the study of the fate of the label as it passes through the tricarboxylic acid cycle. This may be reflected mainly in glutamate, as it is in rapid equilibrium with 2-oxoglutarate. Scheme 1 illustrates the fates of label from the precursors, [1-^{13}C]glucose, [2-^{13}C]glucose, [1-^{13}C]acetate or [2-^{13}C]acetate, in the metabolites observed in our experiments. As ^{13}C label builds up in the glutamate pool as resulting from flux through the cycle, all possible positions of glutamate will be labelled, the positions depending on the precursor. With time the ratio of the singlets to doublets will change and this will correspond to the number of turns through the tricarboxylic acid cycle. An example of the isotopomers of C-3 of glutamate and C-3 of glutamine is given in Figure 5.

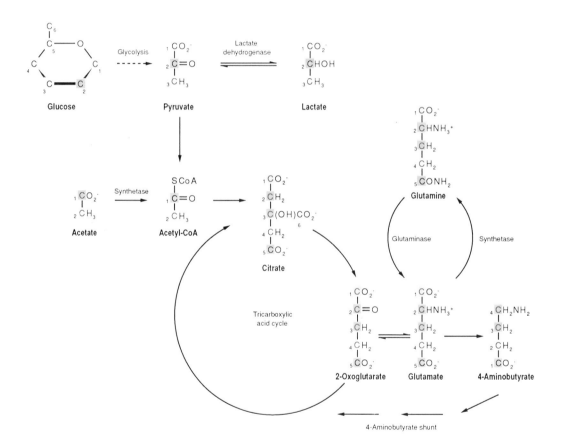

Key: C   initial labelling with [1-^{13}C]glucose or [2-^{13}C]acetate
      C   subsequent labelling with [1-^{13}C]glucose or [2-^{13}C]acetate
      C   initial labelling with [2-^{13}C]glucose or [1-^{13}C]acetate
      C   subsequent labelling with [2-^{13}C]glucose or [1-^{13}C]acetate

Scheme 1. Labelling patterns of the carbon atoms of intermediates from various ^{13}C-labelled precursors[9]

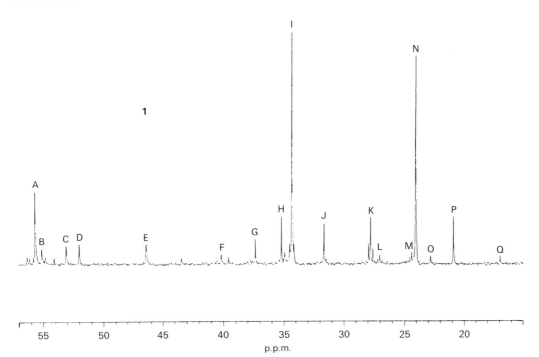

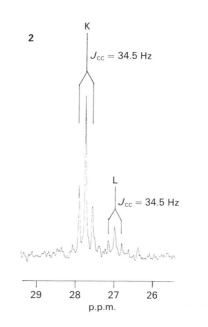

Figure 5. Isotopomers of glutamate and glutamine in a proton-decoupled ^{13}C-n.m.r. spectrum of a brain extract labelled from [1-^{13}C]glucose and [2-^{13}C]acetate

The part of the full spectrum (1) which contains C-3 of glutamate (K) and C-3 of glutamine (L) has been expanded in (2) where the isotopomers are well resolved and can be quantified. A, C-2 of glutamate; B, C-2 of glutamine; C, C-2 of aspartate; D, EDTA; E, C-2 and C-4 of citrate; F, C-4 of 4-aminobutyrate; G, C-3 of aspartate; H, C-2 of 4-aminobutyrate; I, C-4 of glutamate; J, C-4 of glutamine; K, C-3 of glutamate; L, C-3 of glutamine; M, C-3 of 4-aminobutyrate; N, C-2 of acetate, O, -CH$_3$ of N-acetylaspartate (tentative); P, C-3 of lactate; Q, C-3 of alanine. The data are from work in progress in our laboratories.

Using this approach, competing metabolic pathways can be thoroughly dissected. In a ^{13}C-n.m.r. study of the perfused guinea pig heart, Sherry et al.[14] analysed the steady state spectra, i.e. the spectra where changes with time in the intensities of each individual signal in the multiplets no longer occurred. Rat hearts were perfused with [2-^{13}C]acetate, [3-^{13}C]pyruvate or [3-^{13}C]lactate. The multiplet patterns in each of the glutamate resonances changed with each precursor. Figure 6 shows the labelling

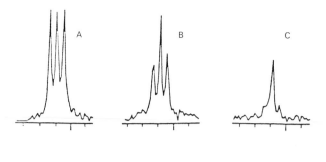

Figure 6. Isotopomers of C-4 of glutamate in a proton-decoupled ^{13}C-n.m.r. spectrum of extracts prepared from guinea pig hearts perfused with [^{13}C]lactate and unlabelled substrates

Analysis of isotopomer patterns shows that less unlabelled precursor is appearing in glutamate in A than in B (perfused also with glucose and insulin), or especially in C (perfused also with acetate). The data are taken with permission from Sherry *et al.*[14].

patterns of the isotopomers of C-4 of glutamate labelled from [3-^{13}C]lactate perfused with different unlabelled substrates: A, 5 mM-[^{13}C]lactate alone; B, 5 mM-[^{13}C]lactate with 5 mM unlabelled glucose plus insulin (10 m-unit/ml), and C, 5 mM-[^{13}C]lactate with unlabelled 2 mM-acetate. In Figure 6, the central resonance is from C-4 of glutamate alone, whereas the two satellite resonances represent glutamate labelled in both the C-3 and C-4 positions. Analyses of these isotopomer patterns enabled the investigators to calculate that the utilization of unlabelled substrates was 30% in A, 45% in B and 68% in C. This led to the conclusion that the guinea pig heart prefers lactate as substrate to glucose, even in the presence of insulin[14].

These studies were extended to the development of a model of the tricarboxylic acid cycle which takes into consideration the relative activities of the net oxidative and anaplerotic pathways[15]. A combination of [2-^{13}C]acetate, which enters the cycle solely through [2-^{13}C]acetyl-CoA (the oxidative pathway) and unlabelled propionate, which enters the cycle via succinyl-CoA (the anaplerotic pathway), was used. The model predicts that if the anaplerotic pathway is inactive, C-2, C-3 and C-4 of glutamate will be equally labelled, and that the ratio of doublet to total enrichment at C-4 of glutamate will reflect the fractional enrichment of acetyl-CoA in the oxidative pathway. The steady-state analysis showed that while pyruvate (as would be expected) competes effectively with acetate for entry to the cycle neither propionate nor glucose with insulin could compete[15].

In a non-steady-state analysis, fractional contributions of different substrates to the total acetyl-CoA pool can be derived, and are independent of number of turns through the tricarboxylic acid cycle or pool sizes[16]. Hearts perfused with a combination of [3-^{13}C]lactate, [1,2-^{13}C]acetate and unlabelled glucose showed the same fractional enrichments at 5 minutes and at 30 minutes: 42% of the acetyl-CoA entering the cycle was derived from acetate, with 32% from lactate and 24% from glucose.

[13]C-n.m.r. isotopomer analysis has also been used to assess the extent of pyruvate recycling (i.e. via malic enzyme or the combined activities of phosphoenolpyruvate carboxykinase and pyruvate kinase) in liver[17] and in brain[18].

INDIRECT [1]H-OBSERVATION OF ^{13}C-LABELLING

As noted in the introduction to this Review, the temporal resolution of ^{13}C-n.m.r. spectroscopy is limited (to about 5–10 minutes) by its low inherent sensitivity. The technique of observation of the protons attached to ^{13}C atoms takes advantage of the far greater n.m.r. sensitivity of protons noted in the introduction, and so enables a theoretical 63-fold increase in sensitivity. In practice, the increase is likely to be less than 30-fold, due to the frequency-dependence of the signal noise. The signal from a proton directly bound to a ^{13}C (unlike that bound to a ^{12}C) is normally split into two resonances due to J-coupling. Use of appropriate spectral editing techniques enables detection of the protons bound to ^{13}C only, and eliminates the signals from all other protons, including those bound to ^{12}C (see [19]). It is important to note that while ^{13}C resonances on -CO_2H and -CH_2- groups can be detected by conventional ^{13}C-n.m.r. spectroscopy, the indirect technique cannot detect the protons on -CO_2H groups because the proton is not directly attached to a ^{13}C. This is a minor limitation in studies on, e.g., amino acid metabolism.

The technique was successfully applied to the calculation of rates of metabolism of lactate and glutamate in rat brain *in vivo*[20]. The same group then used the approach to calculate rates of flux through the tricarboxylic acid cycle in rat brain *in vivo*[21]. They followed rates of flux of ^{13}C from [1-^{13}C]glucose into the C-4 and C-3 positions of glutamate (see Scheme 1). The vastly increased sensitivity of the technique was reflected in their ability to acquire data in less than 98 seconds. The calculated rate of flux through the cycle was approximately 1.4 µmol/min per g, which is in good agreement with previous assessments, and shows clearly that quantitative measurements of metabolic flux can be made *in vivo* in human tissues.

APPLICATION OF ^{13}C-N.M.R. TO CLINICAL DIAGNOSIS

Glycogen synthesis has been studied in gastrocnemius muscles of patients suffering from non-insulin-dependent diabetes mellitus, using ^{13}C-labelled glucose with hyper-glycaemic and hyperinsulinaemic clamps. A 15.5 minute time resolution was achieved and the incorporation of the infused [^{13}C]glucose into the muscle glycogen was significantly lower in patients than in normal control subjects. Glucose uptake and mean rates of non-oxidative glucose metabolism were also found to be lower in patients. The investigators concluded that defects in muscle glycogen synthesis play a dominant role in the insulin resistance occurring in these patients[22].

A major initiative in applying ^{13}C-n.m.r. spectroscopy to clinical problems has been that of Lapidot's group. They have been using a combination of isotopomer analysis of ^{13}C-n.m.r. spectra with g.c.-m.s. to investigate glucose recycling in glycogen storage disease and inherited fructose intolerance in children[11,12].

[U-^{13}C]Glucose was infused naso-gastrically into the children and serial blood samples were taken. The isotopomer patterns of the C-1 and C-2 positions of the glucose enabled them to assess how much of the plasma glucose was being produced endogenously. The studies led them to the conclusion that, in type I glycogen storage disease, the small amount of glucose production was not due to a low rate of gluconeogenesis (as might be expected from lowered glucose 6-phosphatase activity) but that it was more likely to be due to amylo-1,6-glucosidase activity on glycogen. They confirmed that gluconeogenesis was active in type III glycogen storage disease, and also that inherited fructose intolerance arises from an inability to metabolize fructose 1-phosphate (Figure 7). Further analysis of the [^{13}C]glucose isotopomer populations labelled from [U-^{13}C]fructose showed labelling at three adjacent carbon atoms

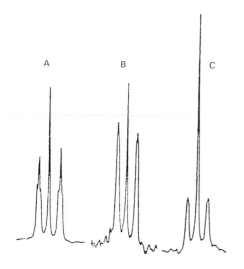

Figure 7. Isotopomers of C-1 of glucose in proton-decoupled ^{13}C-n.m.r. spectra of blood samples from children after infusion with [U-^{13}C]fructose
The patterns show that considerably less unlabelled (endogeneously produced) glucose is appearing in the blood of a patient suffering from inherited fructose intolerance (C) than in a normal child (B). A is a sample of pure glucose containing 3% [U-^{13}C]glucose. The data are taken with permission from Lapidot[11].

on the glucose (^{13}C-3, ^{13}C-4 and ^{13}C-5), suggesting that there may be a direct pathway from fructose to fructose 1,6-bisphosphate which bypasses fructose 1-phosphate aldolase[23].

The excitement of these studies is that they provide a sound basis for essentially non-invasive early diagnosis of many inherited disorders and may detect hitherto unknown metabolic pathways. The techniques briefly described in this Review have only recently been developed and the rate of technical innovation seems certain to result in a rapid expansion of these novel approaches to understanding metabolic regulation in health and disease.

We are grateful to the M.R.C. for financial support, and to Professor P.G. Morris for his constructive comments.

REFERENCES

1. Hevesy, G. (1923) The absorption and translocation of lead by plants: a contribution to the application of the method of radioactive indicators in the investigation of the change of substance in plants. *Biochem. J.* **17**, 439–445
2. Barker, H.A. & Kamen, M.D. (1945) Carbon dioxide utilization in synthesis of acetic acid by *Clostridium thermoaceticum*. *Proc. Natl. Acad. Sci. U.S.A.* **31**, 219–225
3. Canioni, P., Alger, J.R. & Shulman, R.G. (1983) Natural abundance carbon-13 nuclear magnetic resonance spectroscopy of liver and adipose tissue of the living rat. *Biochemistry* **22**, 4974–4980
4. Stevens, A.N., Iles, R.A., Morris, P.G. & Griffiths, J.R. (1982) Detection of glycogen in a glycogen storage disease by ^{13}C nuclear magnetic resonance. *FEBS Lett.* **150**, 498–493
5. Barany, M., Chang, Y.-C. & Arus, C. (1985) Effect of halothane on the natural-abundance ^{13}C n.m.r. spectra of excised rat brain. *Biochemistry* **24**, 7911–7917
6. Cohen, S.M., Rognstad, R., Shulman, R.G. & Katz, J. (1981) A comparison of ^{13}C-NMR and ^{14}C-tracer studies of hepatic metabolism. *J. Biol. Chem.* **256**, 3428–3432
7. Cohen, S.M. (1983) Simultaneous ^{13}C and ^{31}P NMR studies of perfused rat liver — effects of insulin and glucagon, and a ^{13}C NMR assay of free Mg^{2+}. *J.Biol.Chem.* **258**, 14294–14308

8. Van den Berg, C.J. (1973) in *Metabolic Compartmentation in the Brain* (Balazs, R. & Cremer, J., eds.), pp. 137–166, Macmillan, London

9. Badar-Goffer, R.S., Bachelard, H.S. & Morris P.G. (1990) Cerebral metabolism of acetate and glucose studied by ^{13}C-n.m.r. spectroscopy. *Biochem.J.* **266**, 133–139

10. Kalderon, B.,Gopher, A. & Lapidot, A. (1986) Metabolic pathways leading to liver glycogen *in vivo*, studied by GC-MS and NMR. *FEBS Lett.* **204**, 29–32

11. Lapidot, A. (1990) Inherited disorders of carbohydrate metabolism in children studied by ^{13}C-labelled precursors, NMR and GC-MS. *J. Inher. Metab. Dis.* **13**, 466–475

12. Kalderon, B., Korman, S.H., Gutman, A. & Lapidot,A. (1989) Estimation of glucose carbon recycling in children with glycogen storage disease: a ^{13}C NMR study using [U-^{13}C]glucose. *Proc. Natl. Acad. Sci. U.S.A.* **86**, 4690–4694

13. London, R.E. (1988) ^{13}C labeling in studies of metabolic regulation. *Prog. NMR Spectroscopy* **20**, 337–383

14. Sherry, A.D., Nunnally, R.L. & Peshock, R.M. (1985) Metabolic studies of pyruvate- and lactate-perfused guinea pig hearts by ^{13}C NMR: determination of substrate preference by glutamate isotopomer distribution. *J.Biol.Chem.* **260**, 9272–9279

15. Malloy, C.R., Sherry, A.D. & Jeffery, M.H. (1987) Carbon flux through citric acid pathways in perfused heart by ^{13}C-NMR spectroscopy. *FEBS Lett.* **212**, 58–62

16. Malloy, C.R., Thompson, J.R., Jeffery, F.M.H. & Sherry, A.D. (1990) Contribution of exogenous substrates to acetyl-coenzyme A: measurement by carbon-13 NMR under non-steady-state conditions. *Biochemistry* **29**, 6756–6761

17. Cohen, S.M. (1987) Effects of insulin on perfused liver from streptozotocin-diabetic and untreated rats: ^{13}C NMR assay of pyruvate kinase flux. *Biochemistry* **26**, 573–580

18. Cerdan, S., Kunnecke, B. & Seelig, J. (1990) Cerebral metabolism of [1,2-[13]C$_2$] acetate as detected by *in vivo* and *in vitro* ^{13}C NMR *J. Biol. Chem.* **265**, 12916–12926

19. Alger, J.R. & Shulman, R.G. (1984) Metabolic applications of high-resolution ^{13}C nuclear magnetic resonance spectroscopy. *Brit. Med. Bull.* **40**, 160–164

20. Rothman, D.L., Behar, K.L., Hetherington, H.P., den Hollander, J.A., Bendall, M.R., Petroff, O.A.C. & Shulman, R.G. (1985) [1]H-observe/^{13}C-decouple spectroscopic measurements of lactate and glutamate in the rat brain *in vivo*. *Proc. Natl. Acad. Sci. U.S.A.* **82**, 1633–1637

21. Fitzpatrick, S.M., Hetherington, H.P., Behar, K.L. & Shulman, R.G. (1990). The flux from glucose to glutamate in the rat brain *in vivo* as determined by ^{1}H-observed, ^{13}C-edited NMR spectroscopy. *J. Cerebral Blood Flow Metab.* **10**, 170–179

22. Schulman, G.I., Rothman, D.L. & Shulman, R.G. (1990). ^{13}C-NMR studies of glucose disposal in normal and non-insulin-dependent diabetic humans. *Philos. Trans. R. Soc. London Ser. A* **333**, 525–529

23. Gopher, A., Vaisman, N., Mandel, H. & Lapidot, A. (1990) Determination of fructose metabolic pathways in normal and fructose-intolerant children: a[13]C NMR study using [U-^{13}C]fructose. *Proc. Natl. Acad. Sci. U.S.A.* **87**, 5449–5453

Subject index

Page numbers refer to the first and last pages of the relevant *Essay*.